Marburg Virus Disease

Editors

G. A. Martini · R. Siegert

With 131 Figures

Springer-Verlag Berlin Heidelberg GmbH
1971

Editors:

Professor Dr. med. GUSTAV ADOLF MARTINI
Direktor der Medizinischen Universitätsklinik Marburg/Lahn

Professor Dr. med. RUDOLF SIEGERT
Direktor des Hygiene-Instituts der Universität Marburg/Lahn

ISBN 978-3-662-01595-7   ISBN 978-3-662-01593-3 (eBook)
DOI 10.1007/978-3-662-01593-3

© by Springer-Verlag Berlin Heidelberg 1971

Originally published by Springer-Verlag, Berlin · Heidelberg in 1971

Library of Congress Catalog Card Number 72-135966.

Members of the Symposium

ALLEN, A. M., Dr., National Institutes of Health, Bethesda, Maryland 20014, USA

ALMEIDA, J. D., Dr., Royal Postgraduate Medical School, University of London, London W 12, Hammersmith Hospital, Great Britain

ANDERS, W., Prof. Dr., Bundesgesundheitsamt Berlin, 1 Berlin 33, Thielallee 88

AULISIO, C. G., Ph. D., Department of Health, Education and Welfare, Public Health Service, National Institutes of Health, Division of Biologics Standards, Laboratory of Virology and Rickettsiology, Bethesda, Maryland 20014, USA

BALTZER, G., Dr., Medizinische Klinik der Universität Marburg, 355 Marburg, Mannkopffstr. 1, Deutschland

BECHTELSHEIMER, H., Doz. Dr., Pathologisches Institut der Universität Bonn, 53 Bonn 1, Postfach, Deutschland

BEVERIDGE, W. I. B., Prof. Dr., World Health Organization, Geneva, Switzerland

BÖHLE, E., Prof. Dr., Zentrum der Inneren Medizin der Universität, 6 Frankfurt/M., Ludwig-Rehn-Str. 14, Deutschland

BONIN, O., Prof. Dr., Paul-Ehrlich-Institut, 6 Frankfurt/M., Paul-Ehrlich-Str. 44, Deutschland

BORDJOŠKI, M., Dr., Institute for Sera and Vaccines, Belgrade, Yugoslavia

BRÈS, P., Dr., Institut Pasteur, Dakar, Senegal, 36, Avenue Pasteur

CAMAIN, R., Prof. Dr., Institut Pasteur, Dakar, Senegal, 36, Avenue Pasteur

CASALS, J., M. D., Rockefeller Foundation and Yale Arbovirus Research Unit, Yale University School of Medicine, 60 College Street, New Haven, Connecticut 06510, USA

EGBRING, R., Dr., Medizinische Klinik der Universität Marburg, 355 Marburg, Mannkopffstr. 1, Deutschland

ESPAÑA, C. D., Dr., National Center for Primate Biology, University of California, Davis, California, USA

GEDIGK, P., Prof. Dr., Direktor des Pathologischen Instituts der Universität Bonn, 53 Bonn 1, Postfach, Deutschland

GLIGIĆ, A., Dip. Bact., Institute for Sera and Vaccines, Belgrade, Yugoslavia

HAAS, R., Prof. Dr., Direktor des Hygiene-Instituts der Universität Freiburg, 78 Freiburg, Hermann-Herder-Str. 11, Deutschland

HAVEMANN, K., Doz. Dr., Medizinische Klinik der Universität Marburg, 355 Marburg, Mannkopffstr. 1, Deutschland

HENDERSON, B. E., M. D., Virology Section, Microbiology Branch, Laboratory Division, Center for Disease Control, United States Public Health Service, Atlanta, Georgia 30333, USA

HENNESSEN, W., Prof. Dr., Behringwerke, 355 Marburg, Deutschland

HOFMANN, H., Dr., Hygiene-Institut der Universität Wien, A-1095 Wien, Kinderspitalgasse 15, Österreich

JACOB, H., Prof. Dr., Direktor der Universitätsnervenklinik, 355 Marburg, Ortenbergstraße 8, Deutschland

KAFUKO, G. W., Dr., Director of the East African Virus Research Institute, Entebbe, Uganda

KALTER, S. S., Ph. D., Director, Division of Microbiology and Infectious Diseases, Southwest Foundation for Research and Education, San Antonio, Texas 78228, USA

KISSLING, R. E., D.V.M., Virology Section, Microbiology Branch, Laboratory Division, Center for Disease Control, United States Public Health Service, Atlanta, Georgia 30333, USA

KLAŠNJA, R., Dr., University Clinic for Infectious Diseases, Belgrade, Yugoslavia

KORB, G., Prof. Dr., Pathologisches Institut der Universität Marburg, 355 Marburg, Robert-Koch-Str. 5, Deutschland

KUNZ, CH., Doz. Dr., Hygiene-Institut der Universität Wien, A-1095 Wien, Kinderspitalgasse 15, Österreich

LIEBHABER, H., M. D., Department of Epidemiology and Public Health, Yale University, New Haven, Connecticut 06510, USA

MAASS, G., Doz. Dr., Leiter des Instituts für Virusdiagnostik, 44 Münster, Von-Stauffenberg-Str. 36, Deutschland

MALHERBE, H., Dr., Poliomyelitis Research Foundation, P. O. Box 1038, Johannesburg, South Africa

MARTIN, M., Dr., Virology Section, Microbiology Branch, Laboratory Division, Center for Disease Control, United States Public Health Service, Atlanta, Georgia 30333, USA

MARTINI, G. A., Prof. Dr., Direktor der Medizinischen Klinik der Universität Marburg, 355 Marburg, Mannkopffstr. 1, Deutschland

MOCITCH, M., Prof. Dr., University Clinic for Infectious Diseases, Belgrade, Yugoslavia

MÜLLER, G., Dr., Bernhard-Nocht-Institut für Schiffs- und Tropenkrankheiten, Abteilung für Virologie, 2 Hamburg 4, Bernhard-Nocht-Str. 74, Deutschland

NITTNER, K.-R., Dr., Medizinaldirektor, Leiter des Stadt- und Kreisgesundheitsamtes, 355 Marburg, Schwanallee 23, Deutschland

OEHLERT, W., Prof. Dr., Pathologisches Institut der Universität Freiburg, 78 Freiburg, Albertstr. 19, Ludwig-Aschoff-Haus, Deutschland

PETERS, D., Prof. Dr., Bernhard-Nocht-Institut für Schiffs- und Tropenkrankheiten, Abteilung für Virologie, 2 Hamburg 4, Bernhard-Nocht-Str. 74, Deutschland

ROBIN, Y., Dr., Institut Pasteur, Dakar, Senegal, 36, Avenue Pasteur

SCHMIDT, H. A., Dr., Medizinische Klinik der Universität Marburg, 355 Marburg, Mannkopffstr. 1, Deutschland

SCHUMACHER, W., Dr., Ministerialrat im Bundesgesundheitsministerium Bonn, 53 Bonn-Bad Godesberg, Deutschland

SHELOKOV, A., M. D., Prof., University of Texas Medical School at San Antonio, San Antonio, Texas 78229, USA

SIEGERT, R., Prof. Dr., Direktor des Hygiene-Instituts der Universität Marburg, 355 Marburg, Pilgrimstein 2, Deutschland

SIMPSON, D. I. H., Dr., Microbiological Research Establishment, Porton Down, Salisbury, Wilts, Great Britain

SLENCZKA, W., Dr., Hygiene-Institut der Universität Marburg, 355 Marburg, Pilgrimstein 2, Deutschland

SOLCHER, H., Doz., Dr., Psychiatrische und Nervenklinik der Universität Marburg, 355 Marburg, Ortenbergstr. 8, Deutschland

STEFANOVIĆ, Z., Dr., Institute for Sera and Vaccines, Belgrade, Yugoslavia

STILLE, W., Dr., Zentrum der Inneren Medizin der Universität, 6 Frankfurt/M., Ludwig-Rehn-Str. 14, Deutschland

STOJKOVIĆ, LJ., M. D. Sci. D., Institute for Sera and Vaccines, Belgrade, Yugoslavia

STRICKLAND-CHOLMLEY, M., Poliomyelitis Research Foundation, P. O. Box 1038 Johannesburg, South Africa

TAURASO, N. M., M. D., Department of Health, Education and Welfare, Public Health Service, National Institutes of Health, Division of Biologics Standards, Laboratory of Virology and Rickettsiology, Bethesda, Maryland 20014, USA

TODOROVITCH, K., Prof. Dr., University Clinic for Infectious Diseases, Belgrade, Yugoslavia

WATERSON, A. P., M. D., M. R. C. P., Prof., Royal Postgraduate Medical School, University of London, London W 12, Hammersmith Hospital, Great Britain

WILLIAMS, M. C., Dr., Department of Epidemiology and Health, McGill University, Montreal, P. Q., Canada

WOLFF, G., Hygiene-Institut der Universität Marburg, 355 Marburg, Deutschland

WOOD, O. L., Ph. D., Department of Epidemiology and Public Health, Yale University, New Haven, Connecticut 06510, USA.

ZLOTNIK, I., Dr., Microbiological Research Establishment, Porton Down, Salisbury Wilts, Great Britain

Preface

In the late summer of 1967, several patients suffering from a severe disease were admitted to the Department of Medicine of the Marburg University. It soon became obvious that the illness was a hitherto unknown infectious disease. The number of afflicted patients increased to 23. Several cases were observed in Frankfurt/Main at the same time and, some weeks later also in Belgrade, Yugoslavia. Common to all the patients was previous contact with the blood or tissues of Cercopithecus aethiops, the vervet monkey. Altogether 31 people became ill and 7 died.

It was soon apparent that the infectious agent was neither bacterial nor rickettsial in origin but that a viral etiology was probable. Most of the known viral diseases were excluded and the infectious agent was shown to be a hitherto unknown virus with many peculiar characteristics: it infects guinea pigs but not adult mice and is larger than known viruses and of different shape.

This agent was called the "Marburg virus" since most of the cases had occurred in Marburg and the greater part of the laboratory work leading to the detection of the virus was performed in Marburg.

The sudden outbreak of this severe disease, which particularly alarmed all who were concerned with monkeys for whatever purpose, has posed numerous practical, epidemiological and microbiologica' problems, which have occupied the many investigators all over the world. The object of this symposium was to bring these scientists together so that they might present their results and discuss the open questions. Nearly all aspects of the subject were treated: the clinical syndrome and its prognosis, the epidemiology and prophylaxis, the pathology and the microbiology of the disease. The results of the symposium are published in this volume.

We would like to thank all the participants for their valuable contributions, the Ministry of Health of the Federal Republic of Germany and also the Hessian Ministry for Social Affairs for financial support and Dr. Götze of the Springer-Verlag for the publication of this volume.

Marburg, November 1970   G. A. Martini · R. Siegert

Contents

Marburg Virus Disease. Clinical Syndrome

G. A. MARTINI

With 3 Figures

During the last years monkeys from tropical countries, especially green monkeys (certopithecus and macacus), were imported. Originally they were used for pharmacological and physiological experiments. Since it has turned out that kidney cells of monkeys are very convenient for breeding pathogenous viruses, millions of monkeys were needed by virological institutes and pharmaceutic firms throughout the world. As the animals are transported by air directly from their home districts to the institutes it is possible that exotic diseases can be imported—the monkeys can be reservoirs as well as carriers. Until now only single cases of diseases transmitted by monkeys are known as e.g. "herpes simiae" and "Rabies". The "new" disease observed in Marburg, Frankfurt, and Belgrade was the first to occur in an epidemic outbreak.

Own Observations

In 1967, in the middle of August, three patients with the symptoms of an infectious disease were taken to the medical department of the university of Marburg within three days. They had been working in a Marburg factory, where vaccines and sera are produced. The three patients had had direct contact with the green monkeys or their organs. We soon became aware of three more cases in Frankfurt that apparently got the disease at the same time. The circumstances were similar in that all patients had been handling material from monkeys.

Thus after only a few days it became obvious that the disease which was transmitted by monkeys was absolutely unknown in Germany, so that it was necessary to organize emergency public health measures to avoid any spread of the disease. This was very difficult because nothing was known of the infectious agent and the spread of it. In the course of the following three weeks 17 more patients were admitted to our hospital. Two more patients, a female doctor and a nurse, were infected in our hospital. Much later, on November 13th, another patient was admitted who apparently had been infected by her husband. Thus we had 23 patients with this disease and were able to get a survey of the clinical aspects, its different courses and the complications.

In all patients from the Marburg factory a direct contact with blood, parts of the organs or cell cultures taken from the organs could be proved. Ten patients had assisted in the killing or post-mortem examinations of the monkeys; three others had trepanated the skulls. One female patient had dissected the kidneys. Another patient had to handle the cell cultures of the kidneys. Five patients had cleaned the glassware which had contained the cultures. The doctor had hurt herself through the rubber glove with a hypodermic needle, with which she had taken blood from a patient. The nurse had been infected by exposure to the patients.

Incubation Period

The incubation period can be determined in some patients who have only once been exposed to the infectious material (Fig. 1).

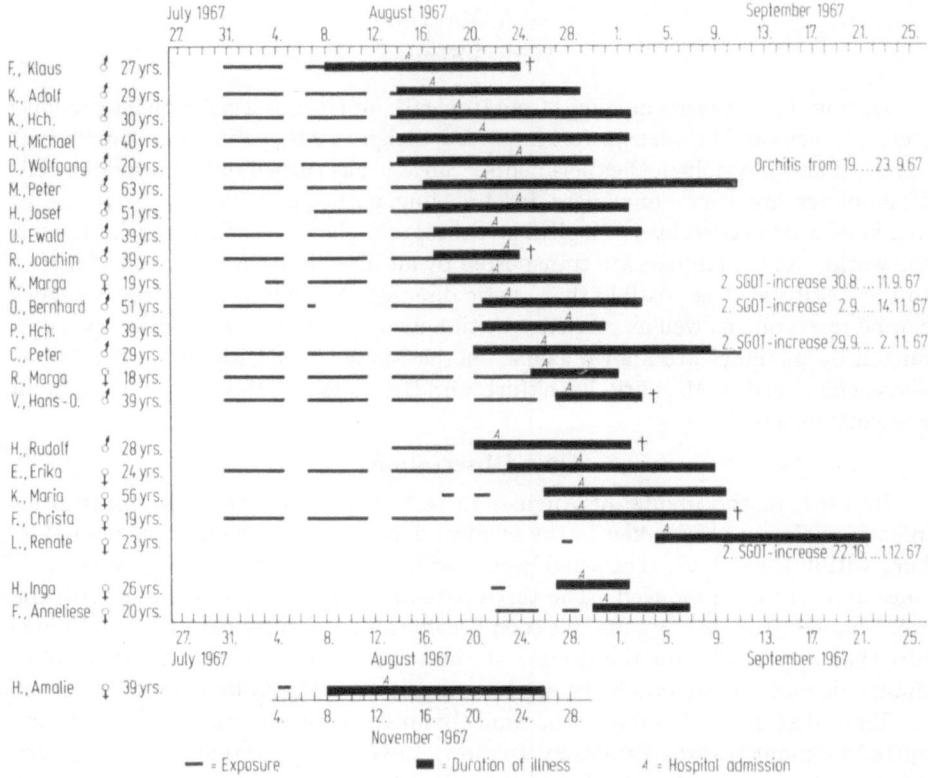

Fig. 1. Time of exposure and duration of illness in 23 patients. Time of exposure is taken as the period during which contact with infectious material was possible

The monkey-keeper HEINRICH P. came back from his holiday on August 13th 1967 and did his job of killing monkeys from August 14th —23rd. The first symptoms appeared on August 21st.

The laboratory assistant RENATE L. broke a test-tube that was to be sterilized, which had contained infected material, on August 28th and fell ill on September 4th 1967.

The only contact of the laboratory assistant MARIA K. with infected material took place on August 18th and 21st. She fell ill on August 26th.

The nurse ANNELIESE K. began to work in the ward on August 23rd and became ill on August 30th.

The doctor INGA H., who was infected by the stitch of a needle on August 22nd, showed the first symptoms of the disease 5 days later on August 27th.

Thus it can be accepted that the incubation period lasted from 5 to 7 days.

Symptomatology

In spite of the different degree of the illness the symptoms and signs were very similar. After having seen the first patients it was not difficult to make the clinical diagnosis. Symptoms and signs were so characteristic that only one misdiagnosis in many suspected cases was made. As an example I should like to give the history of a case which we think to be typical.

Case History

The laboratory assistant MARGA K. had worked since March 1967 in the Marburg serum factory. Her job was to fetch newly removed kidneys from the post-mortem room to clean them and to prepare them for further processing. The prescription was that she had to wear gloves and mouth cover in order to avoid the contamination of the cultures. Occasionally she helped with the fixation of the killed animals on the examination table.

On August 18th 1967 malaise and myalgia took place. On the following day she developed fever up to 39 °C. On August 20th, the third day, for the first time vomiting occurred, one day later burning and reddening of the conjunctiva. The

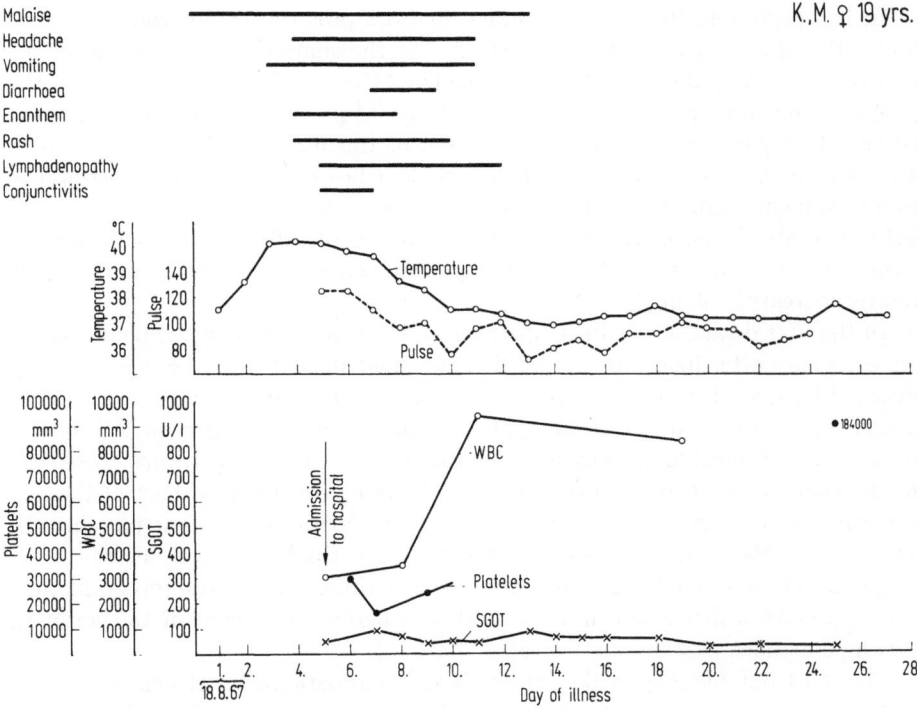

Fig. 2. Main clinical features in a patient whose illness ran a typical course

temperature now rose to 40 °C. On the fifth day she was admitted to our infectious ward. At this time there appeared in the face, on the stem and on the proximal parts of the extremities a macular-papular rash, a red enanthema of the soft palate that reached to the hard palate, and a non-itching erythema of the bigger labias. There

1*

were enlarged lymph-nodes at the neck, along the sternocleidomastoideus and in the axillaries. The liver was not enlarged, the spleen not palpable. There was leucopenia with a shift to the left of the granulocytes and thrombocytopenia. The transaminase was only slightly elevated. Fig. 2 gives a summary of the most important findings.

On the sixth day the conjunctivitis decreased. In the face, the stem and the extremities a maculo-papulous rash developed. The vomiting went on, and a watery, not mucous or bloody diarrhoea occurred, which made an intravenous substitution of fluid and electrolytes necessary. On the eighth day a diffuse cutaneous erythema developed over the whole body. The diarrhoea went on. The vomiting stopped. On the ninth day the state of health improved significantly. The diarrhoea decreased. In the 27th day the skin began to peel off, especially in the face, the palms and the lower extremities. On the 36th day after the beginning of the illness the patient left the hospital. She did not show any more clinical signs, but the reconvalescence was delayed and she continued feeling weak for several weeks.

Summary of the Clinical Symptomatology

In the beginning there was a sudden marked prostration, by which the first day of the illness was clearly to be dated. At the same time a heavy headache occurred, mainly in the area of the forehead and the temples (Fig. 3). Some of the patients also complained of a marked sensitivity of pressure of the eyeballs. Most patients had pains and feeling of tenseness in the muscles of the trunk, most outspoken in the lumbar region. These muscle aches were difficult to distinguish from true meningism, which was seen in two cases only. A lumbar puncture was performed, and in both patients a normal number of cells and in one of them a normal protein content was found. The protein content of the other patient was slightly increased (68 mg%).

In the first days the rise in temperature was accompanied by frequent vomiting, especially after food intake, but also independent from it. The vomiting was preceded by a marked nausea. One or two days later, watery, not bloody and not mucous diarrhoea occurred. Some patients had up to ten voids a day. Whereas the vomiting stopped in most cases at the same time as the temperature decreased, the diarrhoea continued for several days. The more severe the aspect of the disease was, the more marked was the tendency for diarrhoea. In some patients the vomiting and the diarrhoea were preceded by a constipation. Two patients, however, had a chronic constipation during the whole time. Many patients complained of a dry mouth which sometimes occurred even before the advent of the vomiting and the diarrhoea.

The most reliable diagnostic sign was a characteristic rash. It began between the fifth and seventh day at the buttocks, the trunk, and the outside of both upper arms as a distinctly marked, pin-sized dark red papula around the hair roots. This stage lasted up to 24 hours and developed into a macular, papular, sharply delineated lesion which later on coalesced into a more diffuse rash.

Especially in the severe cases, a diffuse dark livid erythema developed in the face, the trunk, and the extremities, which disappeared within several days. During the erythema period some patients had a marked cyanosis. One patient had a herpes labialis before the manifestation of the erythema, another one had several water-clear herpes—like lesions on the abdomen and the first toe, when the rash was disappearing. After the 16th day all patients peeled off, especially on the

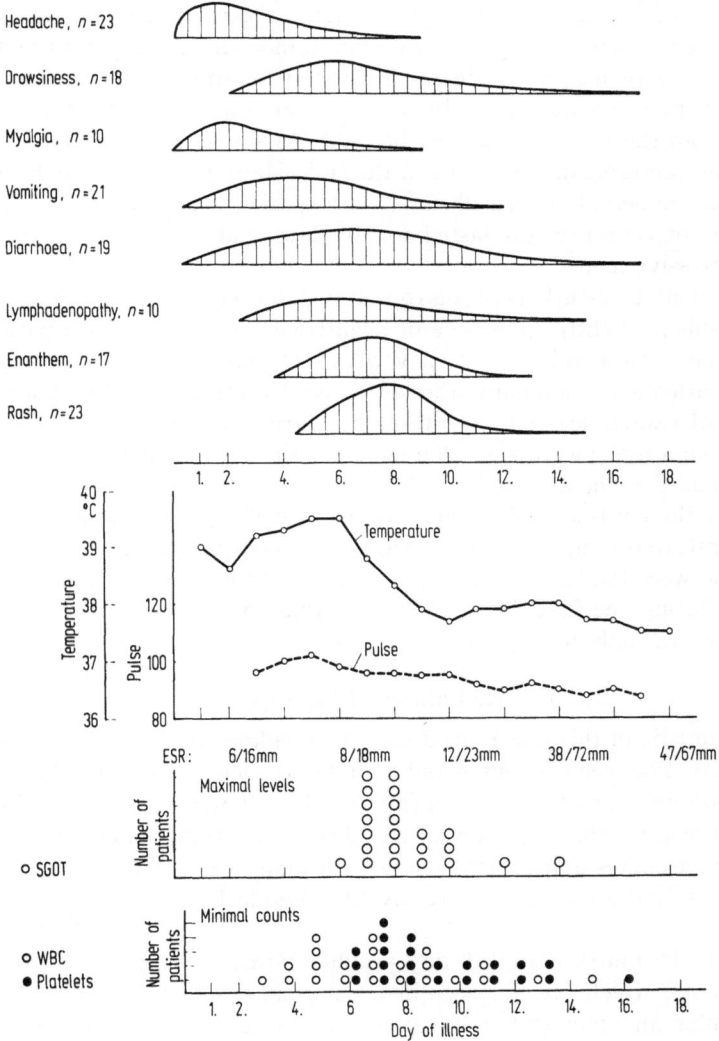

Fig. 3. Incidence and duration of main signs and symptoms: Height of curves in top part is determined by the number of patients with the listed symptom on a given day. For SGOT, WBC and platelet counts, the figure indicates how many patients had reached maximal or minimal levels on the stated day

palms, soles of the feet and the extremities. This lasted from a few days to two weeks. The rash was often accompanied by a scrotal dermatitis or an erythema at the big labias. In about 50% of the patients we saw a conjunctivitis in the beginning or later. The eye fundus did not show any pathological changes, especially no haemorrhage. Nearly all patients who were in the hospital during the decisive period showed a deep dark red colouring of the soft palate, which also spread to the hard palate. This enanthema was much more intensive than the non-specific reddening of the throat, which is often found in feverish infectious diseases. In some patients the enanthema was already seen one day before the development of the rash and disappeared again together with the skin changes. Besides the enanthema there were often fine tapioca-like transparent lesions. In four patients a swelling and reddening of the tonsils with pin-sized, yellowish, not confluent membranes was seen. In some patients there were glandular swellings already before the rash. In four of the fatal courses the patients were very restless, became confused and comatose in the end. Many patients did not remember the most severe period of their illness. In one patient the phase of confusion and impairment of consciousness lasted for 10 days and was followed by anxiety and/or depressive mood.

These mental disturbances disappeared later on. Most of the patients showed a sullen, slightly aggressive or negativistic behaviour. Two patients had paraesthesias with a feeling as if they were lying on crumbs.

Other patients had acroparaesthesias, especially a tingling in the first toe. Two patients had restless legs for several days. During the crisis numerous patients suffered from a hyperaesthesia. One female patient got myelitis after the end of the acute phase of the disease.

Initially there was a rise in temperature above 39 °C. During the first six days the temperature rose up to 40 °C. Then the temperature fell gradually; a second peak was between the 12th and 14th day (Fig. 3). A relative bradykardia occurred, especially during the first days. A tachykardia corresponding to the height of temperature was only found in the fatal cases.

Laboratory Findings

Characteristic of this infectious disease was leukopenia, which was obvious on the first day. Between the fourth and fifth day the leucocyte count partly was as low as 1000/mm³. In all patients a shift to the left was seen in the differential blood picture from the beginning of the illness. In several patients we could also see single myelocytes and promyelocytes. In many cases the granulocytes were changed in a characteristic way, partly like pseudo-Pelger cells or degenerated cells.

Between the fourth and 19th day peculiar forms of lymphocytes were found in all patients, which are not seen in the normal peripheral blood. These were plasmacellular and monocytoid lymphocytes, plasma cells, plasmoblasts, pyroninophil blast cells, or immunoblasts. These plasma cells and similar forms were the most characteristic morphological finding in the peripheral blood. Their total amount in the differential blood picture never exceeded 15%.

In three patients a sternal puncture was performed between the 15th and 17th day. There were a normal erythropoesis and many younger forms of granulocytes.

In two patients the marrow showed an increase of the plasma cells up to 30%. There was a significant increase of non-mature megakaryocytes.

All patients developed an extreme thrombocytopenia. The lowest thrombocyte counts were found between the sixth and twelfth day. Single observations show that the decrease of the thrombocytes began already with the beginning of the disease. In two patients, who died, the thrombocytes decreased to less than 10 000 mm³. In several patients we saw remarkable sludging of the thrombocytes.

During the thrombopenia phase seven patients had a severe haemorrhagic diathesis, leading to haemorrhage from gums, the nose, and puncture lesions as well as to gastrointestinal haemorrhages with haematemesis and melaena. It was remarkable that the rash never became haemorrhagic. Only two patients had purpura-lesions.

In ten patients the Quick prothrombin time, the partial thromboplastin time, the plasma thrombin time, and the fibrinogen were estimated at different times. In none of the patients the plasma coagulation factors were altered so much that the severe haemorrhagic diathesis could be explained by it. Six of 13 fibrinogen examinations showed even slightly increased levels. The total plasma protein decreased significantly in all patients during the illness; in an extreme case to 4.3%. There was no significant proteinuria. A hypoproteinaemia existed also in patients who had no diarrhoea. All plasma fractions were changed nearly to the same extent. Consequently the electrophoresis showed no characteristic changes.

In all patients an increase of SGOT and SGPT was seen. The glutamatdehydrase, the sorbitdehydrogenase, and the glutamyltranspeptidase were also increased in all cases, where these enzymes were examined. The maximum increase of the hepato-specific enzymes was between the 7th and 8th day. In four of the five deceased patients the SGOT increased to 2500 E/l and even up to 5900 E/l. As a rule, the increase of SGPT was lower than the increase of SGOT. The relation between SGOT and SGPT was 7 : 1 in the extreme case. In some patients a second increase of transaminases was seen. This second peak was around the 35th day in three patients. It was interesting that only in one case the increased transaminases were accompanied by an increase of the serum bilirubin in the final phase. The alcaline serumphosphatase was always normal. The creatinphosphokinase was only examined in some severe cases. It was always normal.

In some cases the amylase was significantly increased. Regular controls could not be performed so as not to endanger the laboratory workers and consequently the beginning and duration of the amylase increase could not be determined. One patient had an increase of amylase from the sixth to the 24th day with maximum levels up to 700 Somogyi-units. In 50% we could prove a hypokaliaemia with levels up to 2.3 mval/l. In most cases the hypokaliaemia occurred at the same time as vomiting and diarrhoea. In three cases the hypokaliaemia could not be referred to enteral loss of potassium.

Special Courses

Some patients showed special courses of the disease which were different from the typical course.

Lethal Courses

Of the 23 patients five died between the eighth and 16th day. The age of these patients was between 19 and 39 years.

Special Way of Infection

One patient apparently infected his wife by sexual intercourse. It was proven by the antibody fluorescence method that his sperm contained the infectious material.

Complications

On the twelfth day, after the acute symptoms had disappeared, one patient developed suddenly a bronchopneumonia with pleuritis exsudativa, an increase of temperature, chest pain, yellowish-bloody sputum, and extensive pleural effusion on the left side. From the sputum Staphylococcus aureus and Escherichia coli was grown. After an intensive antibiotic therapy and two pleural punctures with drainage of 800 ml the patient recovered within 6 weeks. This patient and two others developed marked oedema of the lower parts of the legs between the eighth and 16th day, after the total serum protein had decreased below 6 g%. The oedema disappeared within a fortnight after the application of human albumin and digitalis.

A 19-year-old patient, who was only slightly ill and had already left the hospital, got an orchitis on the 36th day after the beginning of the disease. Temperatures rose up to 38 °C, the general condition was nearly normal. The laboratory levels did not show any new changes. The orchitis disappeared five days later after treatment with Prednison. When the patient was examined three weeks later, the left testis was significantly smaller than the right one.

In one patient on the 31st day a new increase of the transaminase up to 113 E/l occurred, without any new complaints. A woman who had been dismissed already and stayed in a reconvalescent home, had again to be admitted to the hospital 73 days after the beginning of the illness because of a new increase of the SGOT up to 200 E/l. The SGPT was 175 E/l. The bromsulphalein retention was 21% after 45 minutes. There were no other clinical or laboratory chemical signs.

The alcaline phosphatase was 25 E/l. The psychic behaviour of this patient had already been very remarkable in the beginning. Now she developed an acute psychosis and had to be transferred to a mental hospital. She later recovered completely.

Therapy

In all patients an antibiotic treatment was tried. Tetracycline, Chloramphenicol, Penicillin, Cephalotin, and Streptomycin, single or in combination, did not influence the course of the disease. In order to avoid a secondary infection, the only treatment left was with Cephalotin. But the temperature did not fall, the diarrhoea did not stop, and the severe courses did not improve.

The therapy of the symptoms was made according to the usual rules. The greatest problem was the therapy of the haemorrhagic complications. As the thrombocytes and the plasma coagulation factors were decreased, a combined treatment with transfusions of fresh blood, concentrations of thrombocytes, fibrinogen, Epsilon-aminocapronic-acid, and vitamin K was performed. Very efficient was the French preparation PPSB from the Centre National de Transfusion Sanguine, Paris, which contains the coagulation factors prothrombin, proconvertin, stuart-factor, and antihaemophil globulin B as a concentration. The severe haemorrhagic tendency could be influenced at least for some hours. Another difficulty was the balancing of the fluid and electrolytes. In nearly all patients electrolytes had to be given, especially potassium. Because of the hypoproteinae-mia several patients got great amounts of human albumin up to 30 g/die. But nevertheless the plasma protein level could sometimes only be kept at 4 g%. In most patients the diuresis worked without any special measures. Some patients were given Mannitol (Osmofundin). In an anuric patient (woman) peritoneal dialysation was performed. Strophanthin and Digitalis were given. Two patients got high doses of Prednison in the final period, but without any success. Very efficient was the treatment with antipyretica, especially of Novalgin (Phenyl-dimethyl-pyrazolon-methylamino-methansulfon acid sodium). In most patients the temperature went down and the general condition improved.

Further Course

Of the 18 surviving patients one woman had a psychosis and another a my-elitis afterwards. The patients were regularly examined after they had left the hospital. All of them complained of an increased exhaustion and sweats and loss of hair. Some patients went on losing weight in spite of good appetite and sufficient intake of calories. Most of them had a tender liver.

A liver biopsy was performed in twelve patients during the reconvalescent period between the fourth and 31st day after the temperature had become normal. Two patients, who at this time still had slightly increased transaminases, showed single cell necroses, accompanied by a not very outspoken fatty degeneration, a slight mesenchymal reaction and focal increase of the Kupffer cells.

The other patients had no characteristic changes or showed a normal morpho-logical picture. In control examinations, the electrocardiogram, the X-ray picture of the thorax, the total protein, the electrophoresis, the SGOT, the SGPT, the serum electrolytes, the blood picture, and the thrombocytes were normal in nearly all patients. An older patient with right bundle branch block in the electrocardio-gram again developed oedema of the lower parts of the legs after having left the hospital and needed a continuous treatment with Digitoxin and Furosemid (Lasix). The total protein and the electrophoresis were normal.

Clinical Course and Prognosis of Marburg Virus ("Green-Monkey") Disease

W. Stille and E. Böhle

With 8 Figures

We present here the main facts about our six patients with what we called "green-monkey disease", whom we observed in August—September 1967 in the Frankfurt/M. University Clinic.

The most important details of the cases are summarized in Table 1 [1]. The main facts are: six cases, five males, one female; four laboratory infections in monkey handlers, two secondary infections in a junior doctor and a pathology assistant. Two early fatal cases, at a time when nobody knew the danger of the illness, two severe but not fatal cases, and two mild infections.

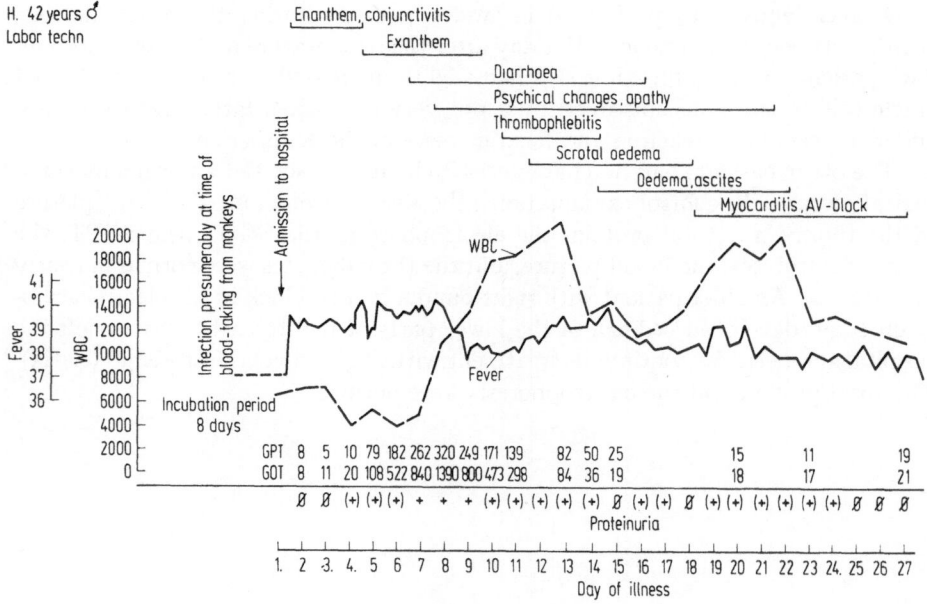

Fig. 1. Course of illness in patient H

The character of the disease as a severe pantropic virus infection involving nearly all systems is better seen in the severe case H. (Fig. 1). It shows the typical "valley and mountain" curve of leucocytosis—the second peak is caused by secondary Klebsiella lung infection and the long—lasting typhoid-like fever which is accompanied by bradycardia. The symptoms of the early phase were exanthema, enanthema and conjunctivitis. The main symptoms of the early organ phase were diarrhoea, slight encephalitis, marked hepatitis with high transaminases, and kidney involvement. The most serious involvement in the late organ phase was myocarditis with typical electrocardiographic alterations and pulse irregularity.

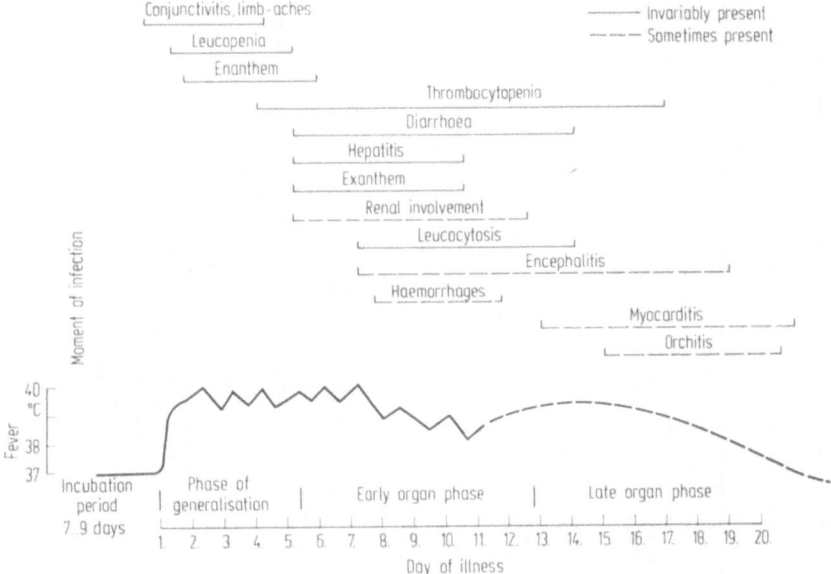

Fig. 2. Scheme of the course of the green-monkey disease

The course of the disease is shown in Fig. 2. From the clinical point of view it is possible to distinguish three phases:
generalisation, early organ phase, late organ phase.

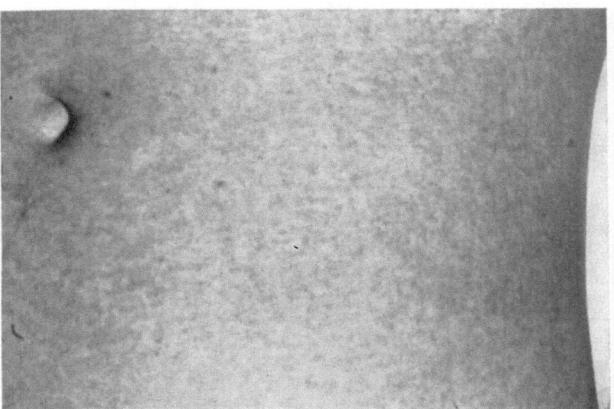

Fig. 3. Exanthem on 7. day of illness

Our opinion about the course of the disease resembles that of Dr. MARTINI and his coworkers in Marburg. There are only small differences. The incubation time of 7—9 days was slightly longer than in Marburg. In our patients there was only a very slight involvement of the lymph nodes.

Table 1. Clinical and laboratory data of the six patients

Patient	Incubation period in days	Duration of illness before hospital admission (in days)	Initial symptoms	Rash	Blood picture	Haemorrhages	Diarrhoea	Liver	Kidney	C.N.S.	Heart and circulation	Fever	E.S.R.	Oedema	Genitals	Course
B. 42-year-old man. Animal attendant	7	7	Fever 1–10th day. Headaches 1–5th day. Conjunctivitis 1–4th day	Scarlatinoid rash from 7th day, later cutaneous haemorrhages	WBC 6400 (7th day) 5200 (8th day) 16,500 (9th day) severe dehydratation (9th day)	Increasing from 8th day, haematemesis, tarry stools; macrohaematuria. Skin haemorrhages	Mild diarrhoea, from day 4, bloody diarrhoea on 7–10th day	3–4 finger-breadths, soft; SGOT 302 mU on 7th day; SGPT 177 mU on 7th day; SGOT 3254 mU on 9th day; SGPT 658 mU on 9th day; Bilirubin 0.82 mg.%	Proteinuria from 9th day. Oliguria on 9–10th day. Urea 20 mg% on 7th day. Urea 79 mg% on 8th day	Clouded consciousness from 4th day; Encephalitis from 7th day; CSF normal	Congestive failure on 9th day. Bradycardia	Continuously from 39.5 to 40.8° C. Agonal defervescence	19/37 on 8th day	Facial swelling	Reddened scrotum	Death on 10th day
P. 64-year-old woman. Animal attendant	7	7	Fever; Clouded consciousness	Scarlatinoid rash from 5th day, later haemorrhagic rash	WBC 23,100 on 7th day 29,650 on 8th day	Macrohaematuria, intravascular haemolysis; haemorrhagic rash	Apparently none	Not palpable	Proteinuria, haematuria. Urea 173 mg% on 8th day	Encephalitis 7th day. Lumbar puncture: 7 cells on 7th day. 24 cells on 8th day	Bradycardia	Continuously about 40° C with a drop in temperature just before death	33/64 on 7th day	None	Unchanged	Death on 9th day
F. 42-year-old man. Animal attendant	7	6	Fever; conjunctivitis; enanthem	Scarlatinoid rash on 6–12th day. Mildly haemorrhagic	Initial leucopenia 41,000 WBC on 6th day leucocytosis of 17,950 on 14th day	Mild petechial bleeding and mucosal bleeding; thrombocytopenia	Severe watery diarrhoea on 6–12th day	Hepatomegaly; anicteric hepatitis. SGPT 150 mU on 10th day; SGOT 375 mU on 10th day. Bilirubin 1.07 mg%. Alkaline phosphatase 427 mU	Proteinuria on 6–17th day. Oliguria on 9th day	Lumbar puncture: 4 cells 7th day. Clouded consciousness. Apathy on 8–15th day	Bradycardia on 16th day, marked STT changes in ECG for the first time — regression within three weeks	Continuously around 39° C with gradual return to normal temperature after 14 days	6/15 on 6th day, 76/105 on 11th day. Slow return to normal	Hypoalbuminaemic oedema and ascites on days 10–15	Scrotal reddening	Cure after long convalescent period

Table 1. Cont.

Patient	Incubation period in days	Duration of illness before hospital admission (in days)	Initial symptoms	Rash	Blood picture	Haemorrhages	Diarrhoea	Liver	Kidney	C.N.S.	Heart and circulation	Fever	E.S.R.	Oedema	Genitals	Course
H. 42-year-old man. Laboratory assistant	8	1	Joint aches; conjunctivitis; enanthem	Scarlatinoid rash on 5–9th day	Leucopenia (2,000 mm³ on 4–6th days) Leucocytosis on 13th day (19,200 pro mm³)	Mild petechial bleeding on 8th day, Rumpel-Leede +, Thrombocytopenia on 8th day	Severe diarrhoea on 7–16th day	Hepatomegaly; anicteric hepatitis. SGOT 1390 mU; SGPT 320 mU; Bilirubin 1.8 mg%. Maximal on 8th day	Proteinuria on 4–24th day. Urea 61.4 mg% on 11th day	Apathy and depression on 7–12th day. Lumbar puncture: 6 cells 18th day	Bradycardia; ST-T segmen abnormality. Myocarditis with A-V block 20th day	Continuously between 39 to 40° C. Biphasic with drop on 10th day	1–14th day normal then increased to 30/60	Hypoalbuminaemic oedema days 15–22	Scrotal pain and reddening; oedema	Cure after long convalescent period
G. 33-year-old man. Morgue attendant	9	1	Tiredness; limb-aches and hyperaesthesia; rash on mucous membranes	Scarlatinoid rash on 5–8th day	Leucopenia of 2,000 pro mm³ on 4th day	Bleeding from gums on 4th day; thrombocytopenia; mucosal rash with slight haemorrhages	Diarrhoea on 4–8th day	Liver palpable (steatosis); mild rise of transaminases; SGOT 67 mU on 5th day; SGPT 56 mU on 5th day	Unremarkable	Headaches; meningism on 14th day. Lumbar puncture: 16 cells 14th day. Restlessness in first phase	Bradycardia	Continuously 2–5th day with second bout of fever on 14th day	17/34 at first. Rising on 7th day to 57/86 on 19th day	None	Scrotal pain Orchitis	Cure after delayed period of convalescence
M. 31-year-old man. Junior doctor	9	2	Backache; achey calves. Hyperaesthesia; conjunctivitis	Scarlatinoid rash on 5–10th day	Leucopenia of 2350 pro mm³ on 5th day	No haemorrhage	Diarrhoea on 5–9th day	Not palpable. Mild rise in transaminases: SGOT 54 mU on 6th day, SGPT 26 mU on 6th day	Unremarkable	Mild somnolescence on 4–6th day	Bradycardia	Persistently between 39–40° C, biphasic with minimal on 8th day	At first normal, followed by mild rise on 15th day	None	Unchanged	Cure after slow convalescence

This table is compiled from earlier publications of the authors: German Med. Monthly 13, 470 (1968) and Dtsch. Med. Wschr. 93, 572 (1968). We thank the publishers for permission to reproduce it here.

The exanthem was more scarlatinoid and showed no involvement of hair follicles. The rash was very similar in all cases (Fig. 3). In the severe cases the exanthem became haemorrhagic; Fig. 4 shows the haemorrhagic skin eruptions

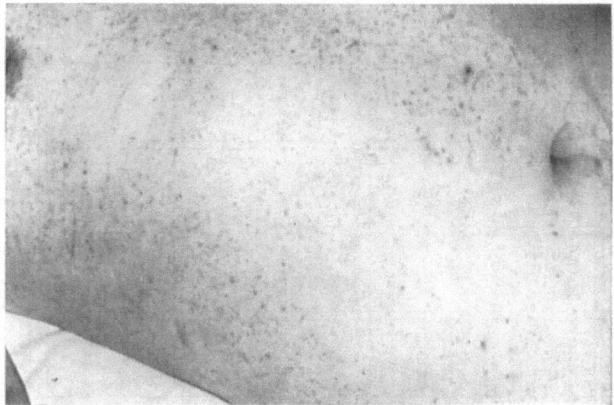

Fig. 4. Haemorrhagic exanthem on 9. day

in our patient B. some hours before his death. The patients had a very typical appearance in the exanthematic phase (Fig. 5). They had a reddened face with some edema, they were uncooperative and morose but not unconscious, except in the

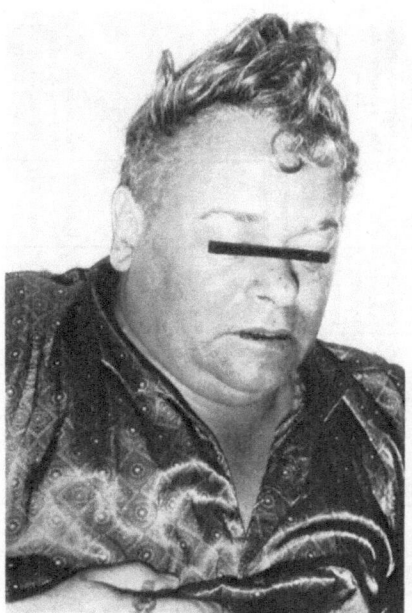

Fig. 5. Patient with severe form of the illness on 11. day

preagonal phase. Figs. 6 and 7, showing one patient in agony, give an impression of the severity of the disease. The patient was unconscious and agitated with marked bleeding from all body apertures.

It was this impression of the disease, together with the signs of serious liver damage, which in the beginning led us to the diagnosis "suspected yellow fever". The two fatal cases died in the early organ phase after only brief treatment in our clinic.

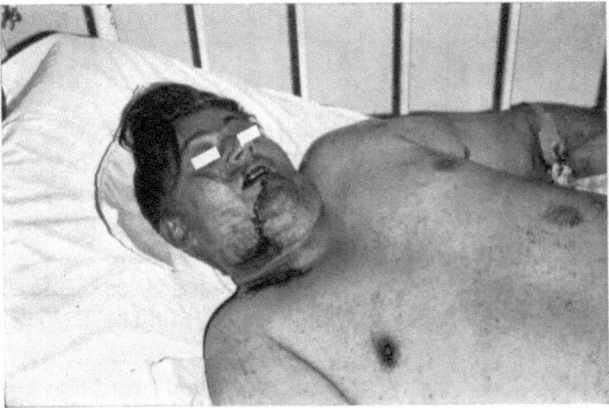

Fig. 6. Patient in the preagonal stage

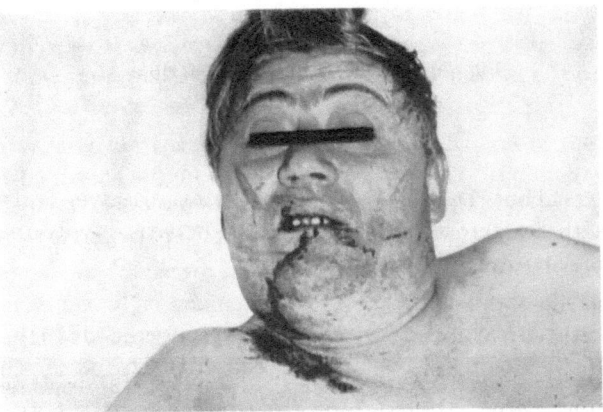

Fig. 7. Patient in the preagonal stage

Therapy

The course of the disease was not influenced by the early application of different antibiotics in high dosage. In the severe cases intensive, symptomatic care was very important; the main measures being maintenance of fluid and electrolyte balance, and circulation therapy.

We have no doubt that these measures were life-saving in our two severe cases. Beside this we were favorably impressed by the use of reconvalescent serum. Our two secondary cases and also the two Belgrade cases were given large quantities of reconvalescent serum, mainly from our first patient F., and in all cases the disease took a mild course without severe visceral manifestations (Fig. 8). Nevertheless, it must be mentioned that in Marburg there were also slight cases where no serum therapy was given.

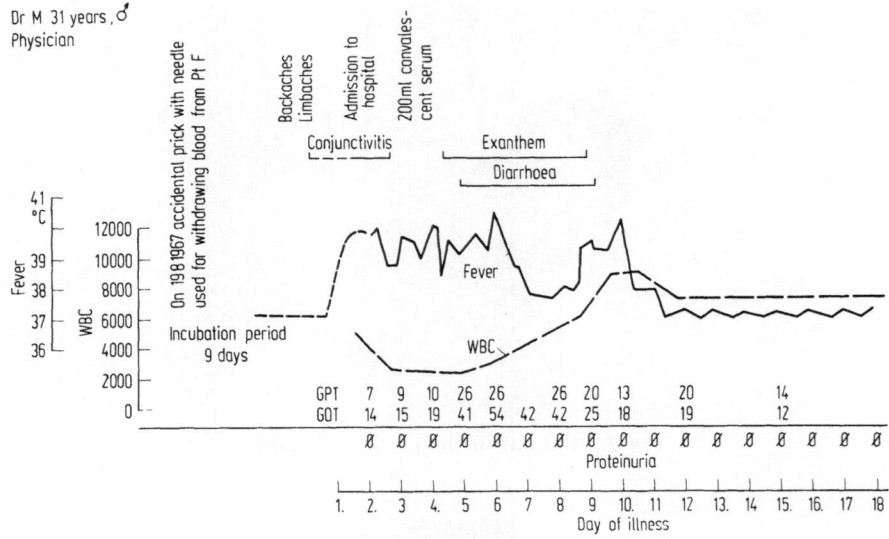

Fig. 8. Course of illness in patient M

Prognosis

The main facts about the prognosis of our patients are given in Table 2. In this table there are three factors of great interest: hepatic involvement, testicular damage, and persistent blood alterations.

Table 2.   Status of patients with "green-monkey disease" in May/June 1969.

	clinical form	liver	testes	virozytes	stenocardia	professiol- ability	MdE
F.	severe	∅	++	+	+	worsened	30%
H.	severe	+	∅	+	∅	worsened	60%
G.	slight	∅	++	+	+	idem	20%
M.	slight	∅	∅	+	∅	idem	20%

We regard as most serious the apparent development of chronic hepatitis in our patient H. After a normalization of the transaminases one month after the beginning of the disease, a second increase of transaminases began three months later (see Table 3). The transaminases became normal again during the second half of last year, but the liver is still enlarged and indurated and bromthalein retention is 11.1%. The clinical picture must now be judged as post-hepatitic

Table 3.   Course of transaminases in patient H.

	before	maximum	15. 9.	27. 10.	7. 11.	31. 1.	4. 5. 68	9. 6. 69
GOT	8	1390	11	16	78	70	31	12
GPT	8	320	17	20	67	82	50	12

Bromthalein 11,1%

cirrhosis. The clinical diagnosis has not been confirmed histologically, because the patient is uncooperative and refuses a liver biopsy. It is uncertain whether the chronic hepatitis is a sequela of the "monkey disease" or of serum hepatitis; the patient had numerous blood and plasma transfusions in the acute stage. A cholecystectomy for cholelithiasis was performed in patient F. one year after the disease. At this time the liver was macroscopically, microscopically, virologically, and clinically normal. The other two patients have no signs of chronic hepatic involvement.

Two patients still have significant and severe testicular damage. One patient has a marked left-sided testicular atrophy, which developed some months after the illness. He shows a normal spermatogram and normal libido, and potency. Another patient has a slight right-sided testicular atrophy, a nearly complete loss of libido and potency, oligospermia but normal ketosteroids. Both patients had suffered from not very severe orchitis in the acute phase of the disease. The two other patients show no signs of testicular damage. We could not observe spermatogenic infections, as described by MARTINI and SCHMIDT [3].

Finally, we want to say something about certain changes in the hemograms of our patients. All patients still have raised blood counts of similar atypical lymphomonocytoid cells, or "virocytes". The exact classification of these cells is difficult — they are regarded mainly as activated lymphocytes. The persisting changes are not so impressive as in the acute disease, but 2—10% of all white cells still show the typical alterations. The hemograms of our four patients in May/June 69 are given in Table 4. We conclude from the persistence of pathological conditions, mainly virocytes, in our patients that chronic and latent infection could also be present.

Table 4.

	F.	H.	M.	G.
Hb in g %	15,2	17,0	14,8	14,8
Ery in Mill	4,84	4,99	4,61	4,58
leucocytes	5700	8100	4800	4450
rod form	8	3		2
segmented form	52	68	53	55
eosinophils		2	2	1
basophils	1		1	
monocytes	6	9	3	9
lymphocytes	28	9	39	23
"virocytes"	5	9	2	10

This result would be in accordance with clinical experience in Marburg, where virus persistence was found in the sperm of the asymptomatic patient [3]. It would also agree with the results of animal experiments. Therefore the patients will require lifelong exact medical supervision, because unexpected clinical developments cannot be excluded. The possibility of new imported infections seems slight. We therefore hope that the Frankfurt/Marburg/Belgrade outbreak will be the first and last reported infection in man.

References

1. STILLE, W., BÖHLE, E., HELM, E., VAN REY, W., SIEDE, W.: An infectious disease transmitted by Cercopithecus aethiops (Green-Monkey-Disease). German Med. Monthly **13,** 470 (1968); Dtsch. Med. Wschr. **93,** 572 (1968).
2. MARTINI, G. A., KNAUFF, H., SCHMIDT, H., MAYER, G., BALTZER, G.: A hitherto unknown infectious disease contracted from monkeys. German Med. Monthly **13,** 457 (1968); Dtsch. Med. Wschr. **93,** 559 (1968).
3. MARTINI, G.A., SCHMIDT, H.: Spermatogene Übertragung des Marburg Virus, Klin. Wschr. **46,** 391 (1968).

Clinical Picture of Two Patients
Infected by the Marburg Vervet Virus

K. Todorovitch, M. Mocitch, and R. Klašnja

With 6 Figures

Nearly at the same time as in Germany — in Marburg/L. and Frankfurt/M. — two patients in Belgrade were infected by the Marburg vervet virus, a veterinarian, Ž. St., 44 years of age, employed in the Institute for Health Protection of the Republic of Serbia, and his wife, 45 years of age.

The epidemiology of their infection directed our attention to the newly imported monkeys *Cercopithecus aethiops*. Our first patient, the veterinarian, besides his other duties, carried out the post-mortem examination of five monkeys. He did not notice any lesion during this time; he performed the post-mortem as usual, with necessary precautions, wearing rubber gloves and other protective garments.

Except for childhood diseases, malaria and suspected, but unconfirmed, gastric ulcer, he was always healthy. His personal and family life was conducted under hygienic conditions. There was no abuse of nutrition or alcohol. No immoderate thirst. Physiological excretions normal.

The disease began on September 1, 1967, with general weakness and raised temperature (38.6 — 39.6 °C) without other clinical symptoms. On the 5th day the temperature rose, the patient shivered, had a headache and on the following two days a heavy diarrhea developed with watery stools not containing mucus or blood. There was general exhaustion, swelling, and erythema of the face, conjunctivitis, a little dry cough but no sneezing. Increasing temperature was followed by dryness of the mouth and the pharyngeal mucous membrane was covered with dry bloodstained mucus, causing difficulty in swallowing and relatively moderate pharyngeal cramps. No sweating. Urination as usual; the urine, especially during the high temperature period, was yellowish. The feces were watery, frequent, and had an offensive odor; he vomited from time to time and had loss of appetite and troubled sleep.

The hematologic examinations while he was still at home showed leucopenia ($L = 2,200$ cmm with 68% polynuclear, without eosionophils). There were no pathologic elements in the yellowish urine, and no pathologic organisms in the feces. Therapeutically, antipyretics and antibiotics (Bemycin and others) were given without noticeable results.

In spite of a strong constitution, the patient, tired and exhausted, presented at his admission a frightened air and the general impression of a very serious illness. The skin was reddish, the mucous membranes of the mouth and pharynx a little swollen, with blood spots. The lips were dry and the conjunctiva yellowish. Sporadic petechiae were seen all over the skin, especially on the face, chest, and

small epigastrium, at some places augmented as larger hematomas. The lymph
glands of the neck were slightly enlarged. Respiration was accelerated ($R = 20$)
without rustles; heart action proportionally retarded ($T = 39.6$ °C, $P = 70$); blood
pressure: 110 : 80. Peristalsis was accelerated, followed by 10—20 watery, bac-
teriologically negative stools daily. Wherever he received injections, there were
both superficial and deep hematomas.

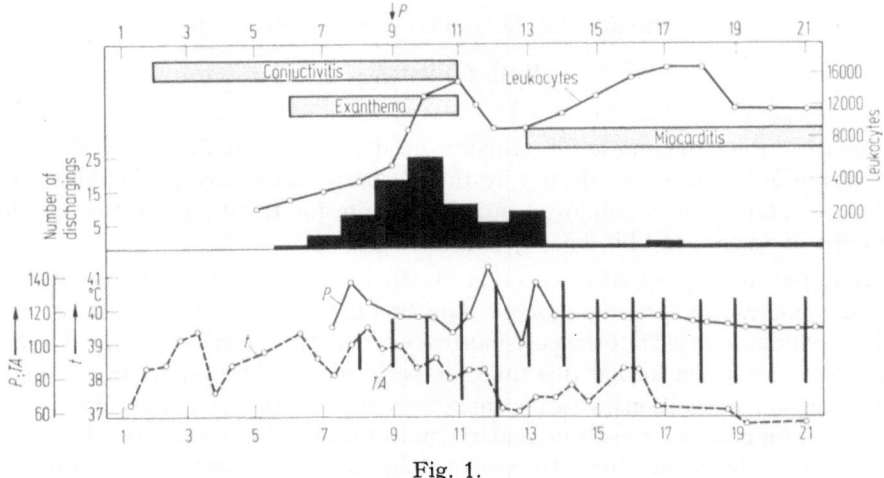

Fig. 1.

On the fifth day of the illness there was leuco- and thrombocytopenia ($L =$
4.000, 6% virocytes; 6% basophils; 52% segmentals; 2% eosinophils; 6% monocytes and 28% lymphocytes; erythrocytes: 4,980,000 cmm; hgl. 95%; SE $= \frac{3}{6}$;
bleeding time $= 2$ min.; coagulation time $= 6$ min. 30 sec.; albuminura, cylin-
druria with granular cylinders, erythruria). (See Figs. 1—3.) The cardinal clinical

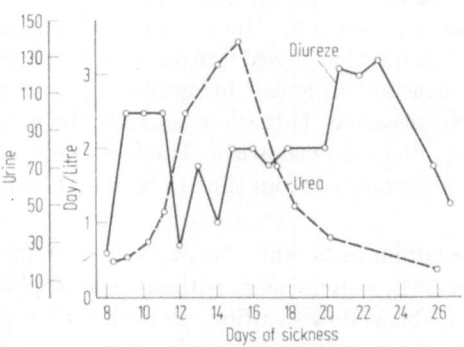

Fig. 2.

symptoms were raised temperature, weakness, headache, insomnia, vomiting,
diarrhea, conjunctivitis, inflamed mucosa, exanthema, hemorrhages in the skin,
impaired function of the liver, kidneys, heart, central and peripheral nervous
system.

An aggravation of the illness made our patient less and less able to move; he was not able to change his position in the bed, to extend his arms, to clench his fist, or to get the spoon to his mouth.

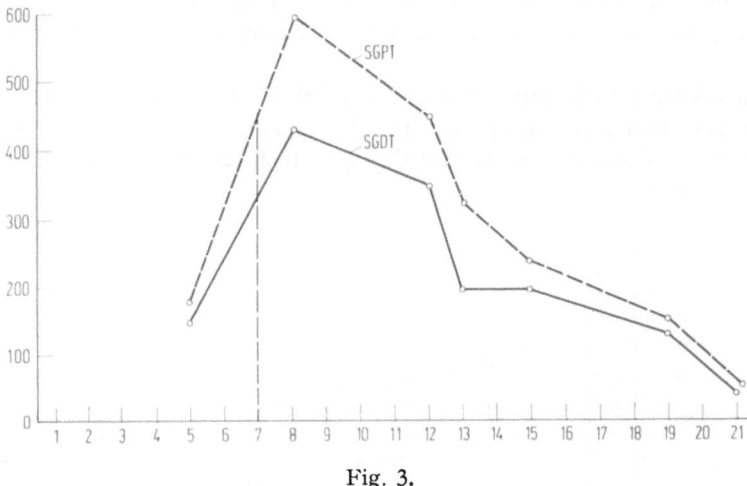

Fig. 3.

Temperature, heart action, function of kidneys and liver, hematological changes, and metabolic states were carefully observed, and after the application of the appropriate perfusions, blood transfusions, antibiotics, cardiacs, and plasma

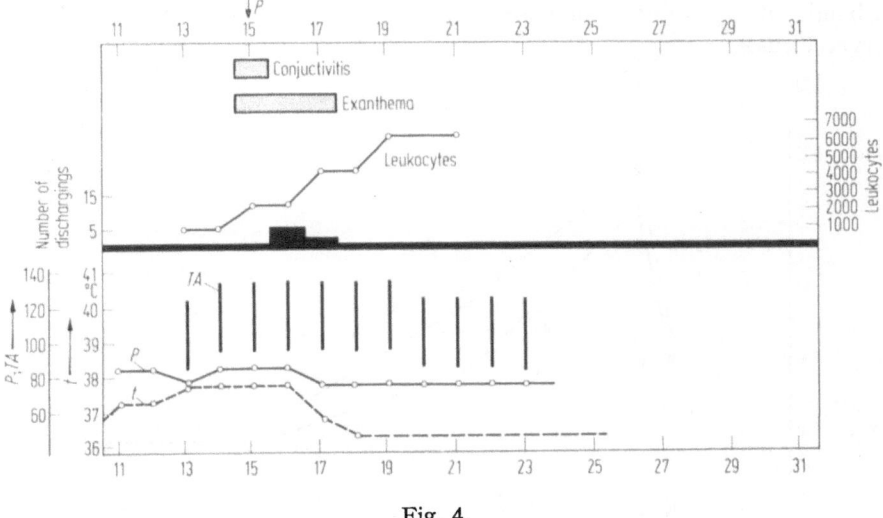

Fig. 4.

and serum infusions from the convalescent German patients, our patient recovered progressively and was able to leave the clinic 32 days after the beginning of his illness, and 25 days after hospitalization.

The second patient, N. St., the wife of the veterinarian, who nursed her husband, fell ill on the 11th day after his sickness began. The first symptoms were:

temperature 38.6 °C, asthenia, headache, pharyngitis, persistent cough, polakiuria, bleeding, diarrhea, pains in the flanks, insomnia, conjunctivitis, skin eruptions.

It is epidemiologically important to notice that the wife had contact with the blood of her husband through the soiled linen on the 4th and 5th days of his illness, when he was still at home. Thus the incubation time may be fixed at 6 to 7 days.

As the following 3 diagrams show (Figs. 4—6), the laboratory findings demonstrate a much milder course of the clinical picture than that presented by her husband. She recovered quickly and was able to leave the clinic after 19 days (13. IX. — 2. X. 67).

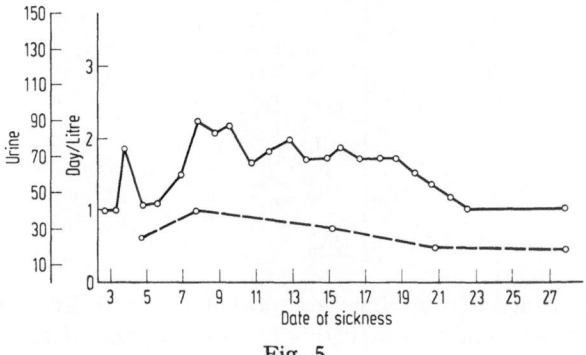

Fig. 5.

As her disease presented the same but considerably milder symptoms than her husband's, it is possible to suppose an infection by a smaller number of Marburg Vervet Viruses.

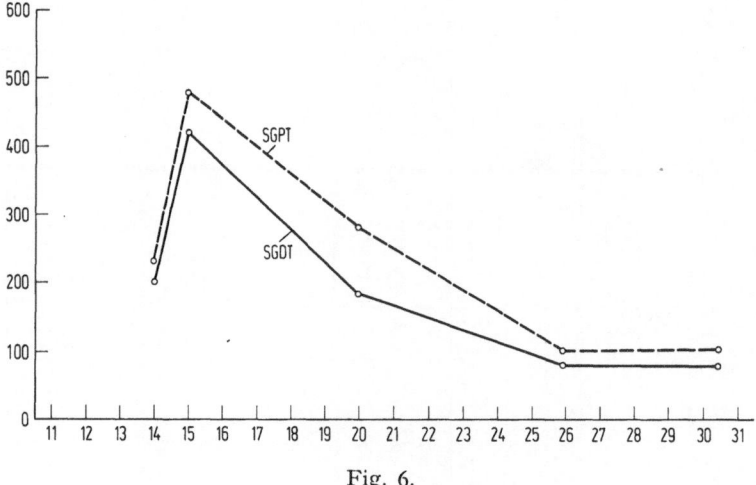

Fig. 6.

The diagnosis of this sickness was confirmed by the isolation the Marburg Vervet Virus through the hemocultures and intraperitoneal inoculations of guinea pigs. These findings were reported to the WHO Center, to Frankfurt, and to Moscow.

As the laboratory researches confirmed dehydration, hemoconcentration, metabolic acidosis, hypocalcaemia, and hypovitaminosis, we applied glycogen, invertose, bicarbonates, lactases, 10% calcium, vitamins, nystatine, durabolin, cardiacs, etc.

Beside transfusions and, the antibiotics, penicillin, extencillin, geomycin, and tetracycline, we obtained very useful therapeutic results by applying plasma and blood serum of the recovered German patients, transferred personally by 2 German experts, Dr. MAY and Dr. BÖHLE, who came to Belgrade in order to learn about the clinical symptomatology and etiology of our two patients.

Summary

A veterinarian and his wife were infected by the Marburg Vervet Virus. Nearly all vital organs were affected among other symptoms; they were given plasma and serum infusions of the recovered German patients. After 32 and 19 days, respectively, they were discharged from hospital. An examination about two years later showed no signs or sequelae of the disease.

Two Cases of Cercopithecus-Monkeys-Associated Haemorrhagic Fever

(Some data on etiology, epidemiology, and epizootology)

Lj. Stojković, M. Bordjoski, A. Gligić, and Ž. Stefanović

With 11 Figures

In August 1967, many cases of a haemorrhagic fever occurred in two laboratories in Germany among laboratory personel handling vervet monkey organs and tissues, or the blood of affected patients.

Many points regarding the etiology and epidemiology of cases have been investigated in laboratories in Germany, Britain, USA, Soviet Union, and Austria [1, 2, 3, 4, 5].

The aim of this report is to present some data regarding the Cercopithecus associated haemorrhagic fever in Belgrade. These data are derived from the limited investigations carried out in our laboratory.

The Source of Infection of Human Cases in Belgrade

Our laboratory has used Cercopithecus monkeys since 1961. The animals were imported in irregular shipments from Uganda through a German dealer. During 1967, most of the shipments went through London airport. From July 18 till August 1, three shipments, each of about 100 animals, arrived in Bel-

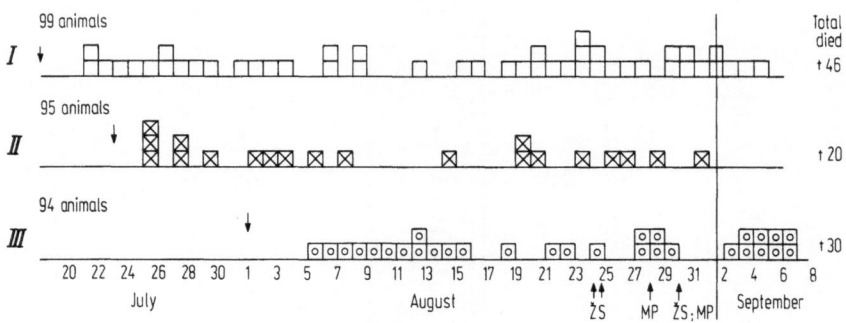

Graph. 1.   Mortality among three shipments of cercopithecus monkeys

grade. The first and the third shipment went through Londen airport, while the second one came to Belgrade, only with a change of plane in München. An unusually high mortality rate was noticed in these three shipments. In Graph 1 data regard-

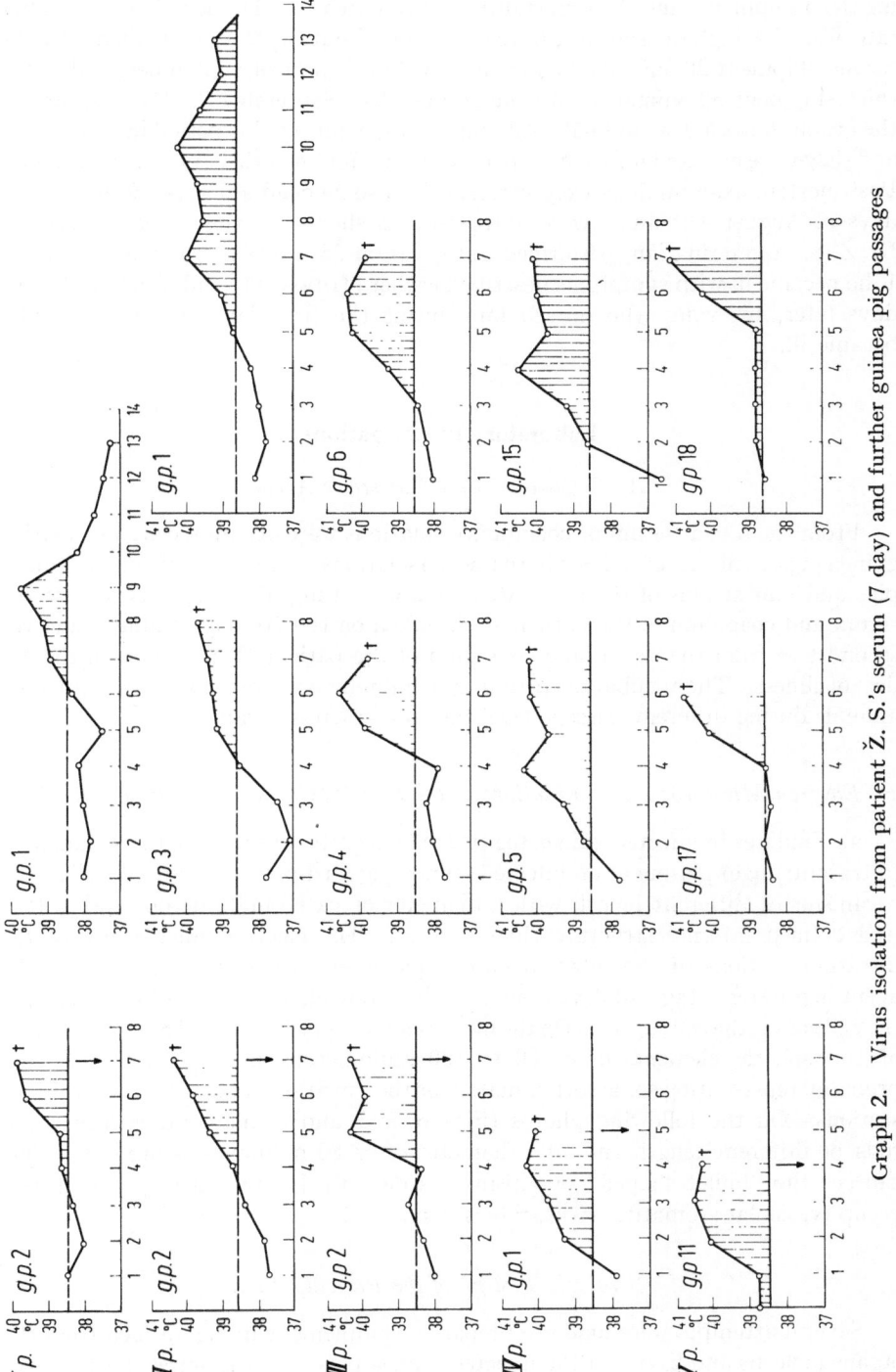

Graph 2.   Virus isolation from patient Ž. S.'s serum (7 day) and further guinea pig passages

ing the 3 shipments and their mortality are presented. In shipment 1 the mortality rate was the highest and in a 6-week period 46 out of 99 animals died. In the second shipment 20 animals died from July 25 till the 1 of September, and in the third shipment 30 animals died from August 1 to September 7. Each square on the graph shows a dead animal. All animals were during this period in quarantine and they were not used for tissue culture work nor for any other type of experiment. Post mortem examinations were performed on some dead animals and in the last days of August 4 animals were autopsied, as shown on the graph by arrows. Dr. Ž. S., a veterinarian, performed autopsies on 25th of August. On September 1, he became ill with symptoms described earlier (Todorović and Mocić [6]). Ten days later, his wife, who nursed him during the first days of his illness also became ill.

Laboratory Investigations

A. Isolations of the Causative Agent

From the blood (serum or coagulum) isolations were attempted by inoculating guinea pigs, adult mice, and newborn mice. Isolations were successful only in guinea pigs and four strains of the agent were established in guinea pigs: two from the serum and coagulum of the patient Ž. S., taken on the 7th day of illness, and the second two from the serum and coagulum of the patient R. S., taken on the 4th day of illness. The incubation period and the temperature curve of the infected animals during different passage levels are presented in Graph 2.

B. Electron Microscopic Examinations of the Infected Guinea Pig Blood and Livers

a) Findings in infected guinea pig plasma: negative staining of the sedimented (ultracentrifuge) plasma of an infected guinea pig (3rd passage) showed rodshaped organisms of different length with a diameter of ca 80 millimicrons and with a well defined internal structure (Figs. 1, 2, 3). The electron microscopy of the ultrathin sections of the infected guinea pig liver showed the presence of the agent in different stages of development. They were all located in the cytoplasm of the cells or on the cell surface. On the microphotographs Figs. 4 and 5, accumulation of the spheric elements of ca 80 to 100 millimicrons in diameter were seen, probably representing some sort of matrix or the "inclusion body" seen in ordinary sections. On the following photos (Figs. 6, 7, 8, and 9) filamentous forms and rods of different length and of a diameter of ca 80 millimicrons are seen. One can see the "bullet shaped" organisms, as described with other agents of this group (vesicular stomatitis virus, rabies, etc.).

C. Serologic Testing of the Patients' Sera

Several attempts were made to prepare a complement fixing antigen from the organs (spleens and livers) of the infected guinea pigs. A satisfactory antigen was prepared from the spleens using a technique of the Microbiological Research Establishments, Porton (personal communication of Dr. Gordon Smith) with

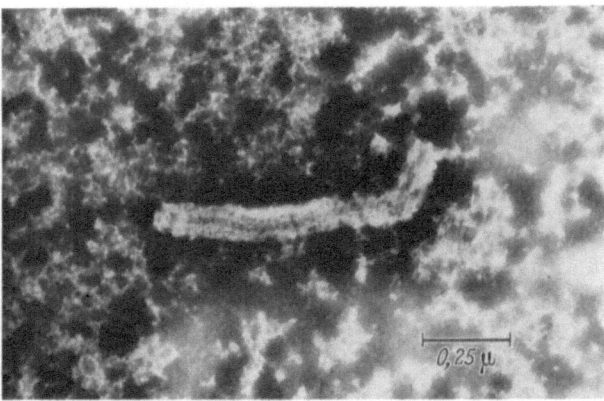

Fig. 1. Negative staining (PTA) of the haemorrhagic fever (vervet associated) virus, found in the blood of an infected guinea pig. Magnification 100 000 × (book size 1 : 2,2)

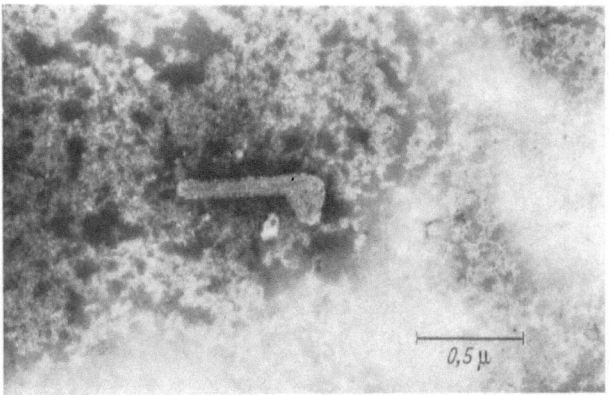

Fig. 2. Negative staining (PTA) of the haemorrhagic fever (vervet associated) virus, found in the blood of an infected guinea pig. Magnification 60 000 × (book size 1 : 2,2)

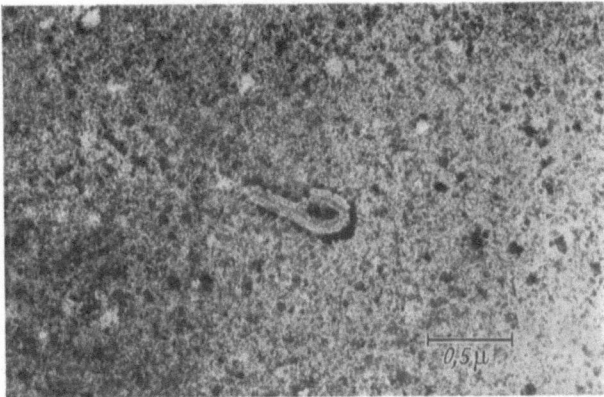

Fig. 3. Negative staining (PTA) of the haemorrhagic fever (vervet associated) virus, found in the blood of an infected guinea pig. Magnification 48 000 × (book size 1 : 2,2)

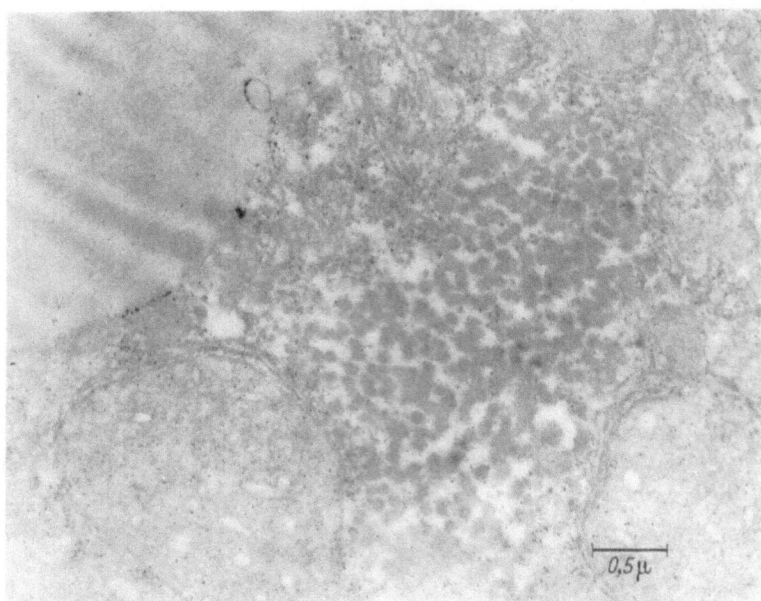

Fig. 4. Part of a cell from infected guinea pig liver showing the accumulation of the early forms. Osmium tetroxide fixation. Araldite (Cargille) embedding. Uranyl acetate. Magnification 45 000 × (book size 1 : 2,2)

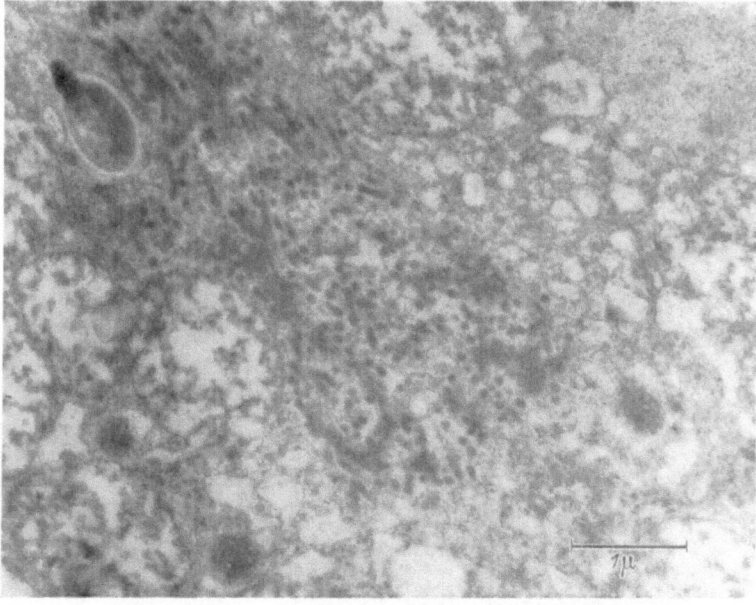

Fig. 5. Part of a cell from infected guinea pig liver showing the accumulation of the early forms. Osmium tetroxade fixation. Araldite (Cargille) embedding. Uranyle acetate. Magnification 35 000 × (book size 1 : 2,2)

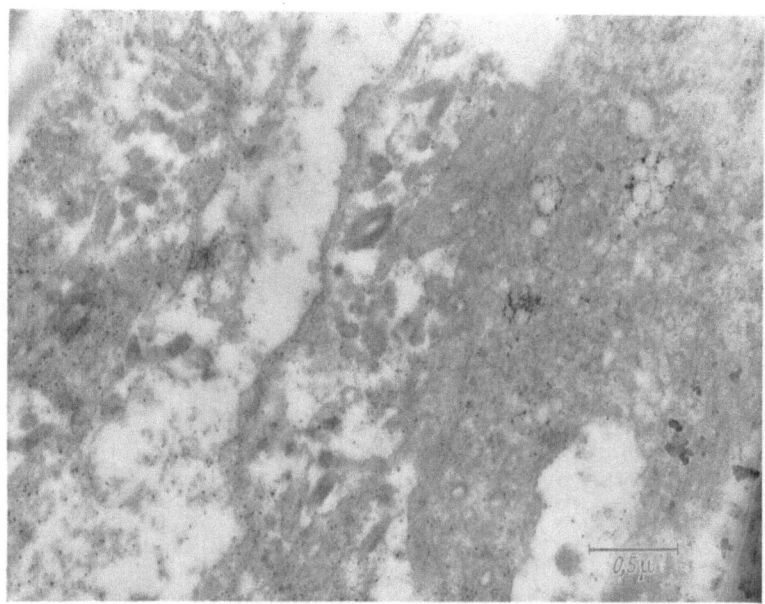

Fig. 6. Parts of the infected guinea pig liver cells showing spherical and "bullet shaped" forms.   Osmium tetroxide fixation, araldite (Cargille) embedding.   Uranyl acetate. Magnification 52 000 × (book size 1 : 2,2)

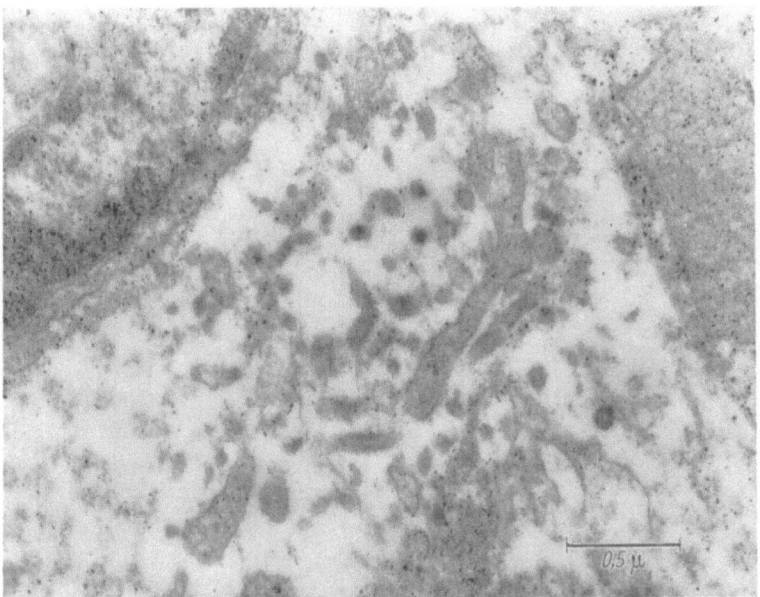

Fig. 7.   Parts of the infected guinea pig liver cells showing spherical and "bullet shaped" forms.   Osmium tetroxide fixation, araldite (Cargille) embedding.   Uranyl acetate.   Magnification 66 500 × (book size 1 : 2,2)

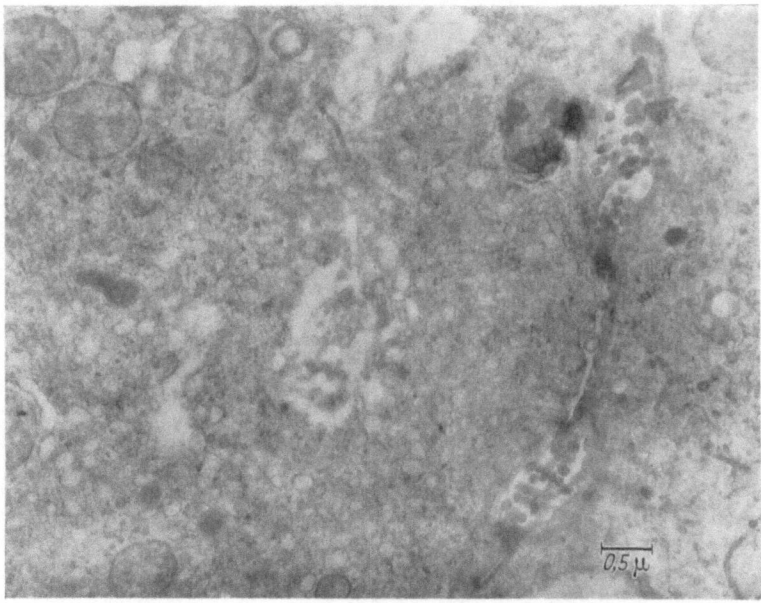

Fig. 8. Part of the infected guinea pig liver cells. Osmium tetroxide fixation, araldite (Cargille) embedding. Uranyl acetate. Magnification 31 500 × (book size 1 : 2,2)

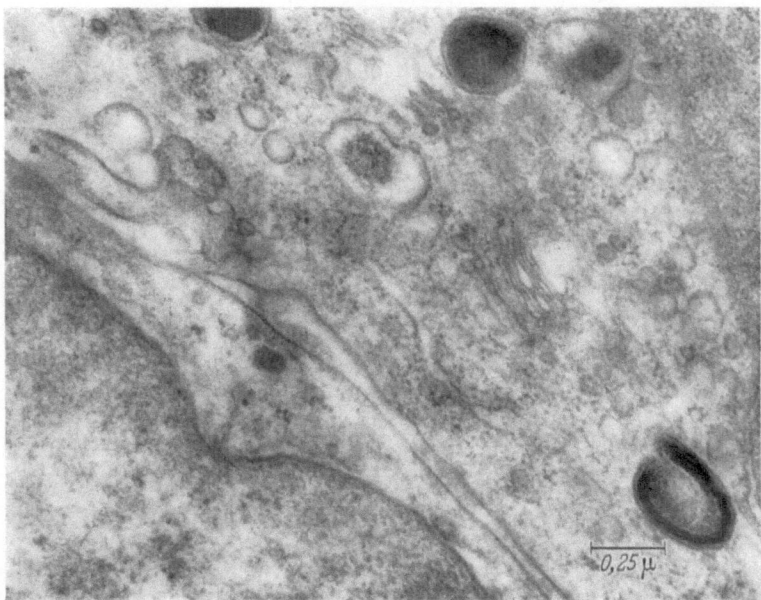

Fig. 9. Part of the cytoplasm of a guinea pig liver cell. Osmium tetroxide fixation, araldite (Cargille) embedding. Uranyl acetate. Magnification 87 500 × (book size 1 : 2,2)

some modifications. The sera were tested with normal antigen as well. Another antigen was prepared from livers, without the addition of inactivating substances.

The following results were obtained:

Patient Ž. S.	Day of illness	Titer (reciprocals)
	7	0
	11	4
	20	32
	30	32
	45	16
	218	8
Patient R. S.	4	0
	11	8
	21	32
	36	16

D. Attempts to Isolate the Agent from the Late Reconvalescent Blood Sample of the Patient Ž. S.

Whole blood taken on 218th day after the onset from the patient Ž. S. was inoculated intraperitoneally into 5 guinea pigs. No febrile reaction was obtained.

E. Agar Gel Precipitation Test with Sera of Patients and with an Antigen Prepared from the Infected Guinea Pig Liver

Attempts to develop an agar gel precipitation test using the antigen prepared from the infected guinea pig liver and the sera of reconvalescent patients were unsuccsessful.

F. The Presence of CF Antibodies in the Sera of Monkeys

a) After the onset of the disease in our patients, 135 animals (Cercopithecus monkeys) were destroyed, and only 57 (from all 3 shipments) were left in quarantine. From September 7 till November 15, only 9 animals died from the remaining 57. The 48 animals in good condition were sacrificed on November 15, and samples of blood and kidneys were taken by aseptic methods.

Complement fixation tests were performed on the sera of the remaining 48 monkeys. Seven tests were not valid due to the anticomplementary action of the serum or to the nonspecific fixation. The results of the testing of 41 monkey sera are given below:

Titer (reciprocals)	No. of animals	%
< 2	5	ca 12%
2	6	
4	4	
8	15	
16 or more	11	ca 88%

Positive reactors were evenly distributed throughout all three groups. All sera were tested in parallel with the antigen prepared from noninfected animal tissues.

As the specific antigen, the same antigens used for testing the sera of patients' were used.

b) The sera of 49 Cercopithecus monkeys taken from the healthy animals imported from KENYA in 1968 and 1969 were tested for the presence of complement fixing antibodies. Only two animals possessed some antibodies in a low titer (1 : 2).

G. Attempts to Isolate the Agent from the Kidneys of 30 Animals Originating from the in 1967 Infected Lots

The kidneys from each of the 30 animals, sacrificed on November 15, 1967 were pooled in 6 lots, according to their serologic status, and a 10 percent suspension was inoculated into groups of 5 guinea pigs. All inoculated guinea pigs remained without fever for a period of 12 days. A second passage was performed with the pool of the blood of all guinea pigs of one group, and again all animals were symptom-free.

Discussion

Two cases of a haemorrhagic fever occurring in Belgrade in August 1967 were part of the same episode which occurred in Marburg and Frankfurt. The low incidence in Belgrade was due to the fact that the infected monkeys were kept in quarantine and were not used for any work. The unusually high death rate exceeding the earlier death rates was most probably due to the existing infection. The infection of Dr. Ž. S. who performed post mortem examinations occurred on 25th of August, most probably through some small abrasions on the unprotected forearm or through the conjunctivae. His wife R. S. was infected most probably by contact with a cotton swab soaked with the blood of the patient.

Further spread of cases was also stopped thanks to excellent cooperation by our colleagues from Germany who warned us in time to take all necessary measures in the laboratory and in the hospital.

Isolations in guinea pigs and electron microscopic examinations showed the same findings as with Marburg agent, and the morphology of the agent as well as its location within the cell suggested that both our agent and the Marburg agent belonged to the same group, designated as Rhabdovirus group (MELNICK) [7].

The complement fixing antibodies in our patients showed the same or lower titers in comparison with the tests on the same sera performed in Porton (G. SMITH) and in Moscow (CHUMAKOV, personal communication). But these tests confirmed that our patients possessed antibodies reacting with an antigen prepared from the agents isolated in Germany.

The results of the CF tests performed on the vervet monkeys coming from infected lots of animals showed a high proportion of positive reactors. We consider these results as specific, due to the fact that the same antigen was used for testing the patients' sera, which reacted in a very specific manner. On the other hand, sera from vervet monkeys collected in 1968 from the other lots were without antibodies. Our impression was that some unspecific reactions were more frequently observed with antigen prepared from livers, and that the spleens were more suitable for antigen preparation.

Our findings suggest that the monkeys were infected originally in Africa, and that the temporary holding in London did not infect the animals.

We were not able to find any virus in the kidneys of monkeys coming from the infected lots. The presence of antibodies and the failure to isolare virus from healthy animals might indicate that this disease in Cercopithecus monkeys is not a latent virus infection.

References

1. SIEGERT, R., HSIU-LU SHU, R., SLENCZKA, W.: Isolierung und Identifizierung des ,,Marburg-Virus". Deutsche Med. Wschr., 93, 604 (1968).
2. GORDON SMITH, C. E., SIMPSON, D. I. M., BOWEN, E. T. W., ZLOTNIK, I.: Fatal human disease from vervet monkeys. Lancet, 1967 11.
3. KISSLING, R. E., ROBINSON, R. Q., MURPHY, F. A., WHITFIELD, S. G.: Agent of disease contracted from green monkeys. Science 160, 888 (1968).
4. CHUMAKOV, M. P. and coworkers: Personal communication.
5. HOFMANN, H., KUNZ, CH.: Komplementbildende Antikörper nach Infektion mit dem ,,Marburg-Virus" (Rhabdovirus simiae) beim Menschen. Zbl. Bakt. I. Abt. Orig., 229, 288 (1969).
6. TODOROVITCH, K., MOCIC, M., Belgrad: Clinical picture of 2 Patients Infected with Marburg Vervet Virus. This Symposium.
7. MELNICK, J. L., McCOMBS, R. M.: Classification and Nomenclature of Animal Viruses, In: Progress in Medical Virology 8 (1966). Edited by J. L. MELNICK, Basel—New York: S. Karger.

Haematological Findings in Marburg Virus Disease: Evidence for Involvement of the Immunological System

K. Havemann and H. A. Schmidt

With 5 Figures

Haematological changes in viral disease are well known. Thus the "Marburg-virus" transmitted by monkeys and recently observed in 23 patients of the Marburg area [1, 2] also provoked a typical haematological response. The particular haematological features studied in these patients may justify special consideration.

10 of the 23 patients exhibited a moderate, generalized, enlargement of lymph nodes in the course of the disease, predominantly between the third and the sixth day of illness. The slightly enlarged lymph nodes were indolent, soft, and of medium size. In all 5 patients who died, a moderate swelling of the hilar and abdominal lymph nodes were noted [3]. A palpable spleen has only been observed in one patient, whereas autopsy revealed an enlargement of the spleen in 2 of the 5 casualities.

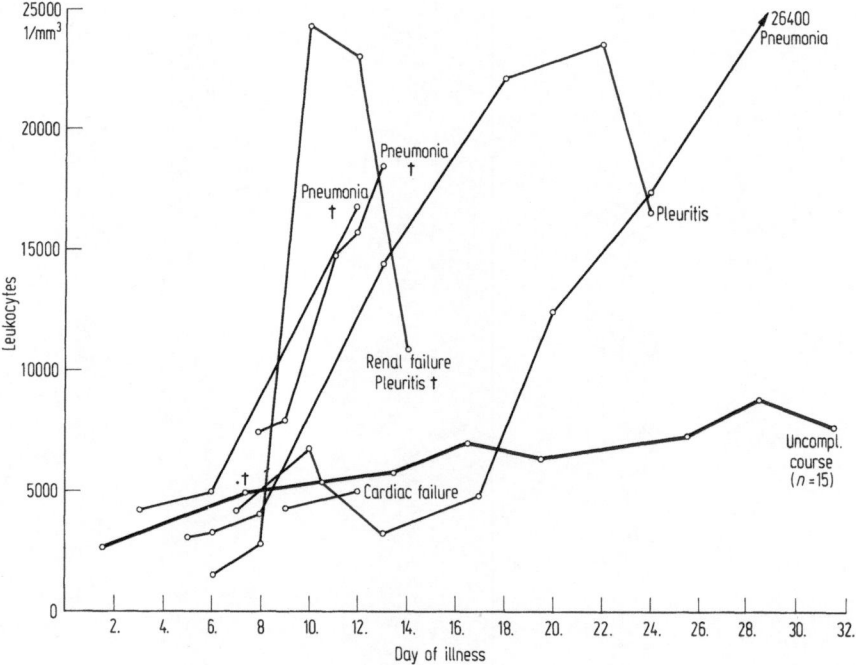

Fig. 1. Mean values of total white cell counts of 16 patients with uncomplicated course and white cells of single patients with complicated course who finally died. Each dot represents mean values of a 3 day's period

Thrombocytopenia complicated by bleeding was seen in several patients and will be discussed in a special paper on this symposium (see R. EGBRING et al.).

The most characteristic haematological finding was a leucocytopenia, which was already detectable at the first day of the outbreak and which was most pronounced between the second and the fifth day of illness (Fig. 1). A this time in some patients the number of leucocytes fell below 1000/mm³. In patients *without complications* the leucocyte-count increased to 5000—6000 mm³ during the second week of illness and later on showed values of 8000 and more. In the course of *complications* such as bronchopneumonia and pleuritis a constant and pronounced leucocytosis developed.

Fig. 2 shows the mean values of absolute counts of different leucocytes. In this curve the values of all the patients were included except for those with a complicated course and those who died. In order to demonstrate the abnormality of the differential leucocyte count, the counts were related to normal values [4]. It is unknown, however, up to which point the patients had a normal blood count. The

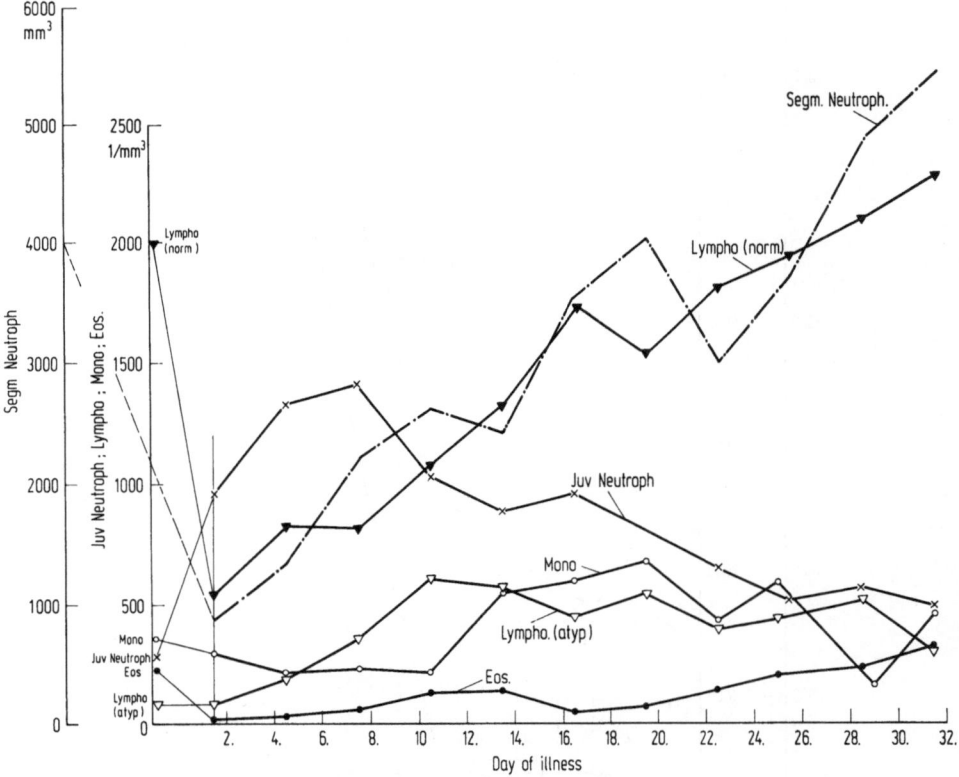

Fig. 2. Course of mean values of the specified white cells of 16 patients

abnormality of the blood counts seen in the first days of illness indicate that abnormalities were already present during the incubation period. During the first week of ill health the absolute count of neutrophils, lymphocytes, and eosinophils was diminished. Furthermore, from the beginning the differential count in each case showed a considerable shift to the left with some metamyelocytes and only

3*

a few myelocytes. The shift to the left was most pronounced towards the end of the first week. It should be emphasized that during the different phases, i.e. the period of leucocytopenia and the later stages, the relation of neutrophils to lymphocytes did not differ from a normal differential count.

The second week of illness was characterized by a gradual increase of leucocytes in which all types of white cells were included. During the same period the shift to the left was reversed. At the end of the first week and the beginning of the second an increase of normal-looking lymphocytes appeared together with atypical lymphocytes.

Whereas the *absolute* number of atypical lymphocytes remained constant throughout the 4 weeks of observation, the *relative* amount decreased while neutrophils and normal lymphocytes rose to normal values.

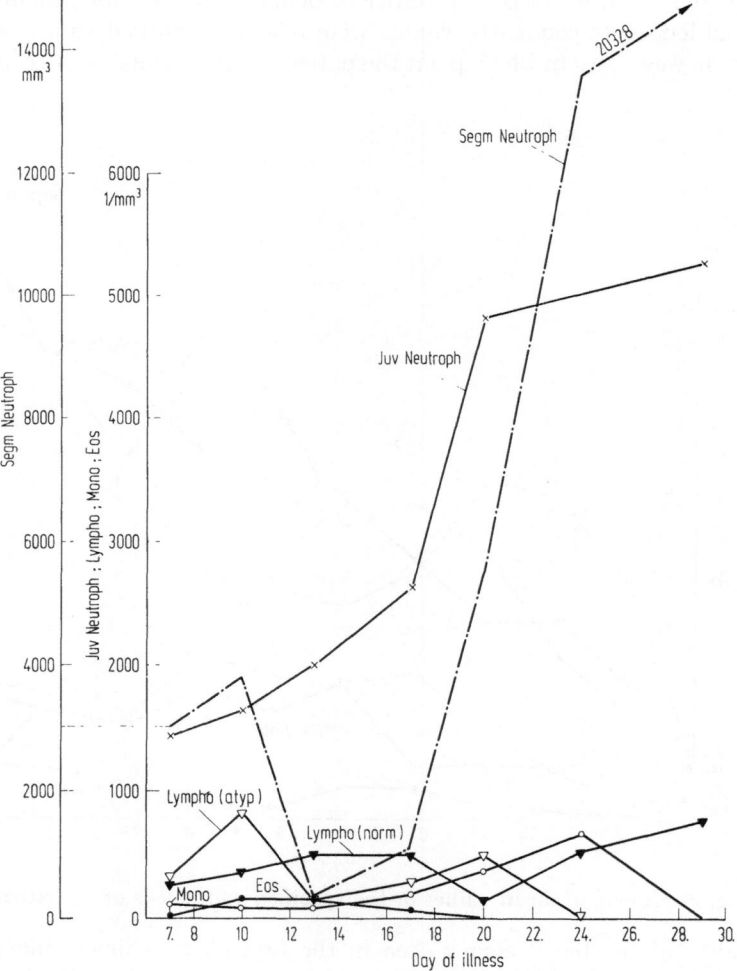

Fig. 3. Behaviour of total white cells during the course of the disease in patient K., Heinrich. Juvenile neutrophils and other white cells present in peripheral blood. 12 days after all acute signs of the disease had subsided he developed bronchopneumonia and exudative pleurisy.

In the third week a monocytosis was seen with moderate eosinophilia and later on an exceedingly pronounced lymphocytosis.

In contrast to the findings of the uncomplicated course, the differential count of the complicated disease developed along different lines. These were caused mainly by bacterial superinfections and were followed by excessive leucocytosis, pronounced shift to the left and eosinopenia during the acute phase of the infection (Fig. 3).

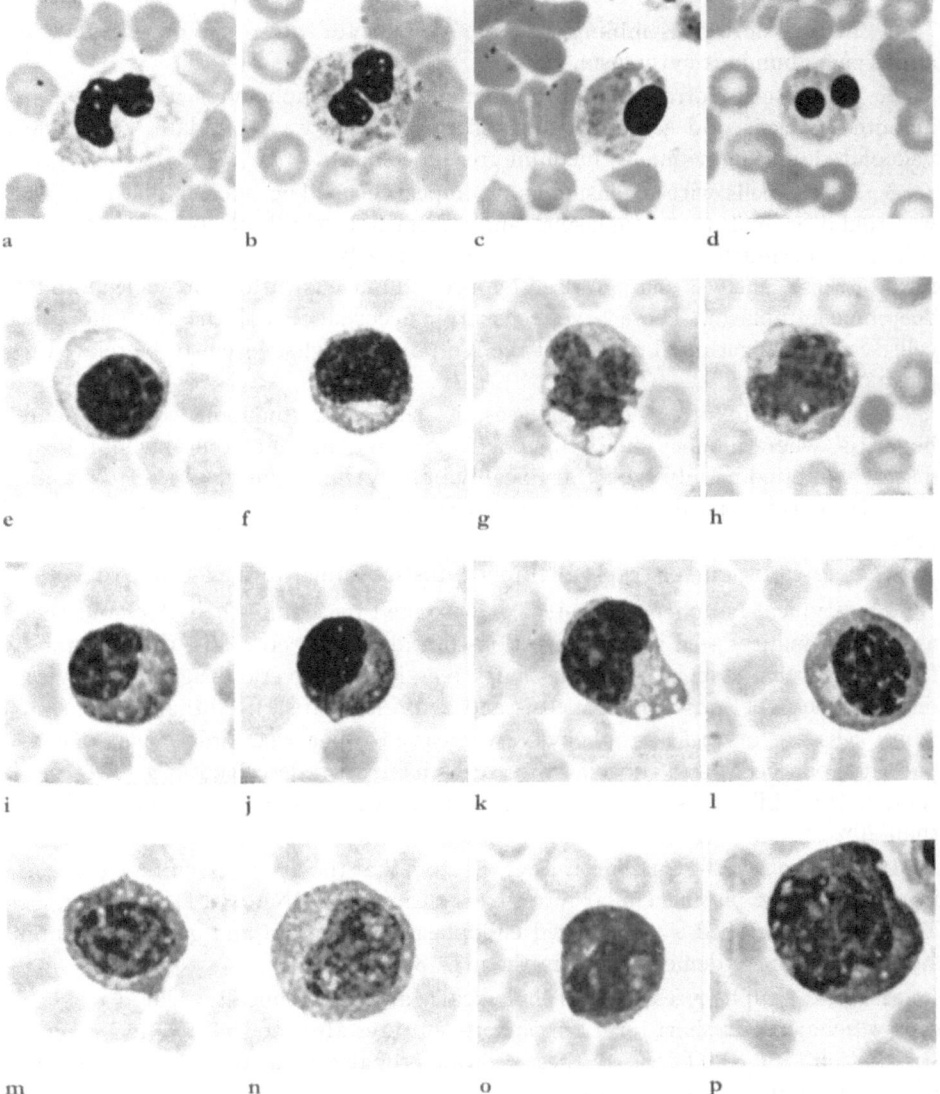

Fig. 4. Cells in peripheral blood of patients. a), b) Pelger cells; c), d) disintegrating granulocytes with pyknosis; e) large lymphocyte; f), g), h) lymphoid monocytes; i), j), k) plasma cells; l) immature plasma cell; m), n), o), p) pyroninophilic blast cells (immunoblasts)

In addition to these quantitative haematological alterations several specific qualitative changes occurred. During the phase of leucocytopenia and the appearance of immature granulocytes, degenerated neutrophils were seen quite often. These cells exhibited a pycnotic or fragmented nucleus and a vacuolized cytoplasm. Sometimes they had a pseudo-Pelger cell appearance (Fig. 4).

In addition the above-mentioned atypical lymphocytes were seen. With reference to the classification of Wood [5] we were able to differentiate between 3 types (Fig. 4).

1. Type A closely resembling the large lymphocyte with round or oval nucleus and a pale abundant cytoplasm.

2. Type B cells corresponding to the monocytoid lymphocytes with an oval or indented nucleus and some nucleoli. The cytoplasm was abundant, sometimes vacuolated, and light blue to medium grey in colour.

3. Type C cells with a plasmacytoid appearance. The nucleus of these cells was round or oval, the cytoplasma ranges from light sky blue to intense blue in colour. A perinuclear clear zone is present in nearly all cells. A more immature form of these cells was characterized by a round nucleus with some nucleoli and a deep blue cytoplasm. These larger cells are in accordance with the pyroninophilic cells seen in the germinal centers of the lymph nodes which seem to be related to plasmoblasts.

In contrast to other viral diseases, i.e. infectious mononucleosis or acute hepatitis, where monocytoid lymphocytes are predominant, the Marburg Virus disease exhibited mainly type C atypical lymphocyte and this is related to plasma cells. Moreover, typical plasma cells with excentric nucleus and plasmoblasts were seen.

Therefore this kind of atypical lymphocytosis is similar to that seen in rubeolae.

The atypical lymphocytes were first described by Türk [6] in 1898 and they are predominantly seen in viral infections but also in tuberculosis, the posttransfusion syndrome, and in hypersensitivity reactions to several drugs [5]. In viral disease, like infectious mononucleosis and acute hepatitis, more than 20% of the white cells are atypical lymphocytes whereas in other virus infections such as parotitis, varicella, and rubeolae these cells have been described in a lower percentage [5]. Like these, the Marburg Virus Disease also showed counts of less than 20%.

It has been found by numerous investigators that the atypical lymphocyte, and in particular the plasmacellular form, has an increased RNA and DNA synthesis. These cells developed a pronounced endoplasmic reticulum and gamma-globulin synthesis has been demonstrated in them [7, 8, 9].

Atypical lymphocytes are morphologically and biochemically related to blast cells which occur in short term lymphocyte cultures after nonspecific and specific stimulation [5, 7, 9]. Therefore these cells are probably transformed lymphocytes, which were induced by the virus infection.

According to the findings of Hall [10] and Crowther [11] atypical lymphocytes originate from the lymph nodes involved during the peak of primary and secondary immune reactions and enter the peripheral blood by the efferent lymph. They probably are responsible for the distribution of the proliferative immune response throughout the whole lymphatic system. Post mortem examinations of patients with the Marburg Virus disease showed a predominance of this cell type in spleen and lymph nodes [3].

Examinations of the bone marrow were done only in a few patients. In one of the aspirates of the marrow we found 21% plasma cells during the beginning of atypical lymphocytosis (Fig. 5), whereas in smears of the later stages 5 to 10% were counted only.

As compared to the granulocytopenia and thrombocytopenia in the peripheral blood, a marked increase in promyelocytes and myelocytes and in immature megakaryocytes of the bone marrow occurred (Fig. 5).

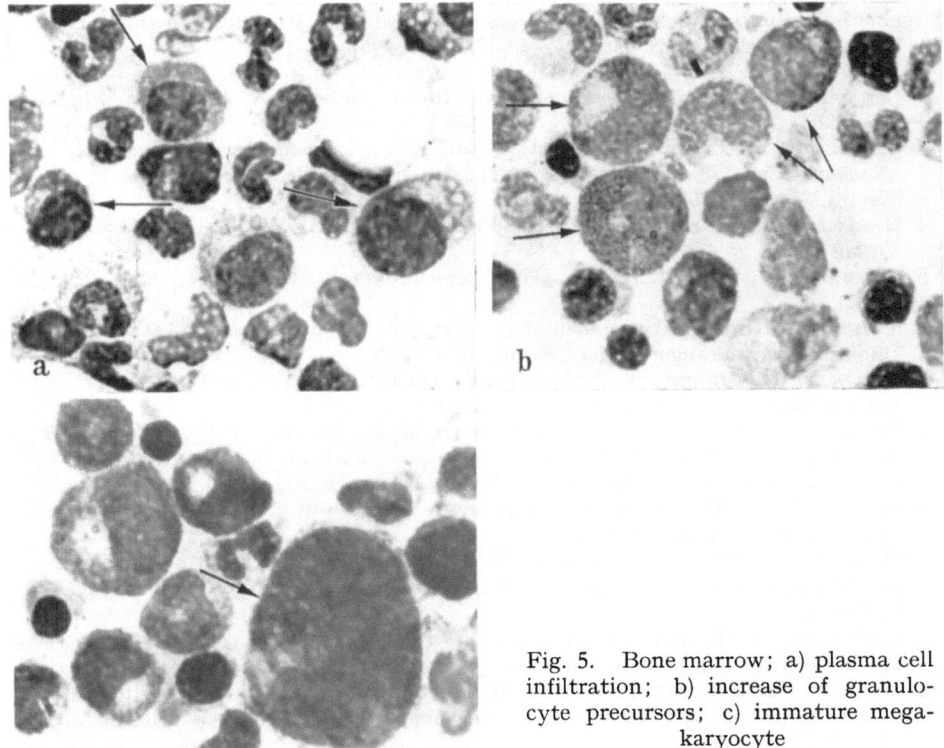

Fig. 5. Bone marrow; a) plasma cell infiltration; b) increase of granulocyte precursors; c) immature megakaryocyte

There is some evidence that the degeneration seen in neutrophils of the peripheral blood is caused by the virus directly and that the neutropenia is due to histiocytic phagocytosis of degenerated cells. This assumption is in accordance with the appearance of basophilic bodies in histiocytes of the gastrointestinal tract and other tissues which may be phagocytized neutrophils [3, 12].

In summary, the following typical haematological findings were seen in Marburg Virus disease:

1. An initial leucocytopenia in which all white cells are involved.

2. The development of degenerated neutrophils.

Initial leucocytopenia and degenerated leucocytes could be observed constantly before the typical clinical symptomatology developed.

3. The disease with uncomplicated course showed a normal ratio of granulocytes to lymphocytes during all stages. However, when complications developed, excessive leucocytosis and shift to the left were seen — evidence for bacterial superinfection.

4. Atypical lymphocytes, predominantly of a plasmacellular type, appeared simultaneously with an increase of plasma cells in the bone marrow. These findings are related to a marked proliferative response of the immunological system similar to that of other viral diseases.

References

1. Martini, G. A., Knauff, H. G., Schmidt, H. A., Mayer, G., Baltzer, G.: Über eine bisher unbekannte, von Affen eingeschleppte Infektionskrankheit: Marburg-Virus-Krankheit. Dtsch. Med. Wschr. **93**, 559 (1968).
2. Stille, W., Böhle, E., Helm, E., van Rey, W., Siede, W.: Über eine durch Cercopithecus aethiops übertragene Infektionskrankheit. Dtsch. Med. Wschr. **93**, 572 (1968).
3. Gedigk, P., Bechtelsheimer, H., Korb, G.: Die pathologische Anatomie der „Marburg-Virus"-Krankheit (sog. „Marburger Affenkrankheit"). Dtsch. Med. Wschr. **93**, 590 (1968).
4. Begemann, H., Harwerth, H.-G.: Praktische Hämatologic. Stuttgart: Thieme Verlag 1969.
5. Wood, T. A., Frenkel, E. P.: The atypical lymphocyte. Am. J. Med. **42**, 923 (1967).
6. Türk, W.: Klinische Untersuchungen über das Verhalten des Blutes bei akuten Infektionserkrankungen. Wien: Braumüller 1898.
7. Inman, D. R., Cooper, E. H.: The relation of ultrastructure to DNA synthesis in human leucocytes. Acta haemat. **33**, 257 (1965).
8. Bond, V. P., Fliedner, T. M., Cronkite, E. P., Rubini, J. R., Brecher, G., Schork, P. K.: Proliferative potentials of bone marrow and blood cells studied by in vitro uptake of H^3 thymidine. Acta haemat. **21**, 1 (1959).
9. Chessin, L. N.: The circulating lymphocyte — its role in infectious mononucleosis. Ann. Int. Med. **69**, 333 (1968).
10. Hall, I. G., Morris, B., Moreno, G. D., Bessis, M. D.: The ultrastructure and function of the cells in lymph following antigenic stimulation. J. exp. Med. **125**, 91 (1967).
11. Crowther, D., Fairley, G. H., Sewell, R. L.: Lymphoid cellular responses in the blood after immunisation in man. J. exp. Med. **129**, 849 (1969).
12. Haas, R., Maas, G., Oehlert, W.: Untersuchungen zur Tierpathogenität eines von Cecopithecus aethiops übertragenen menschenpathogenen Erregers. Med. Klin. **63**, 1359 (1968).

Clinical Manifestations and Mechanism
of the Haemorrhagic Diathesis in Marburg Virus Disease

R. Egbring, W. Slenczka, and G. Baltzer

Technical Assistance G. Wolff

With 1 Figure

In the course of the Marburg virus disease 7 out of 23 patients developed a severe haemorrhagic diathesis. Clinical manifestations consisted of haemorrhages from the mucosa of the mouth and nose, as well as of the whole intestinal tract. In some cases petechial bleedings of varying severity were also observed in the skin. In most patients protracted bleeding from injection areas of the skin was noticed.

Table 1 and 2 show a compilation of all the macroscopically and microscopically determined haemorrhages in five of the patients who succumbed to the disease. In addition in all patients a more or less pronounced thrombocytopenia existed which was considered to be the cause of the haemorrhagic diathesis. In those cases where the disease took a serious course the haemorrhagic diathesis was quite obviously also the cause of the lethal outcome.

Table 1. *Sites of bleeding (macroscopic) in five patients with Marburg virus disease who died.* (Gedigk et al. 1968)

	R., J.	V., O.	H., R.	F., C.	F., K.
Mucosa of the mouth and nose	+	+	−	+	+
Gastrointestinal mucosa	+	+	+	+	+
Spreading intracutanous bleedings at injection site	+	−	−	−	+
Subcut. and intracut. bleeedings	+	−	−	+	−
Renal pelvis	−	−	−	+	−
Subepicardial and endocardial bleedings	−	+	+	−	+
Lung parenchyma	−	+	+	+	+
Conjunctival bleedings	−	−	+	−	−
Omentum majus	−	−	+	−	−

In one of the patients[1] it was possible at the peak of the illness to demonstrate a severe degree of coagulopathy with a reduction of factor V and II as well as a deficiency in fibrinogen. The thrombin time was prolonged to more than 4 minutes.

---

[1] We have to express our thanks to Dr. Schwick (Behring Werke) for the investigation and interpretation of this case.

This proves an inhibition of coagulation by fibrinogen degradation products which can result from increased primary or secondary fibrinolysis or fibrinogenolysis. In some of the other patients prolonged thrombin times and cephalin times were also noted, which also indicates a coagulopathy in these cases.

Table 2. *Microscopic bleedings in five patients with Marburg virus disease who died.* (Gedigk et al. 1968)

	R., J.	V., O.	H., R.	F., C.	F., K.
Gastric mucosa	—	+	—	—	—
Myocardium of the right ventricle	—	+	—	—	—
Interstitial tissue of the testes	—	—	+	—	+
Lymph nodules	—	—	+	—	+
„Perifollikuläre Ringblutungen"	+	—	—	—	—

In table 3 (taken from McKay and Margetten 1967) viral diseases are listed, in the course of which occasionally severe haemorrhages were observed, which could have an unfavourable effect on the illness. The cause of the haemorrhagic diathesis was, according to McKay and Margetten, a so-called disseminated

Table 3. *Virus diseases with occasional severe bleedings complicating the course.* (McKay and Margetten, 1967)

A. Exanthematous virus diseases
1. Varicella
2. Vaccinia
3. Variola
4. Rubella
5. Rubeola

B. Arbovirus diseases
1. Thai haemorrhagic fever
2. Bolivian haemorrhagic fever
3. Philippine haemorrhagic fever
4. Argentine haemorrhagic fever
6. Kyasanur forest disease

generalized intravascular clotting. Platelet and fibrinous thrombi, respectively, were found in the arterioles and capillaries and ischaemic necroses in nearly all organs. Tables 4 and 5 give a survey of the microscopical findings which show varying degrees of severity. Gagel et al. recently showed that in fowl pest, another virus disease, a severe degree of coagulopathy is also seen.

The disseminated intravascular coagulation which leads to a consumption coagulopathy can be caused by various mechanisms, for instance, by the breakdown of tissue and endothelial or blood cells as well as thrombocytes. This also applies to the effect of bacterial endotoxins and anoxaemia of long duration.

As a result of the associated and increased fibrinolysis, fibrinogen degradation leads to the formation of antithrombin VI, which prolongs the thrombin time. In addition, during intravascular coagulation, monomer complexes of fibrin

Table 4. *Pathologic changes in "certain virus infections", complicated by severe haemorrhage.* (McKay and Margetten, 1967)

Platelet and/or fibrin thrombi associated with haemorrhage or ischemic necrosis of the organ involved.

Liver: 1. No changes
      2. Sinusoidal platelet or fibrin thrombi
      3. Focal necrosis
      4. Extensive infarction

Kidney:
      1. No changes
      2. "Lower nephron nephrosis, or acute focal tubular necrosis"
      3. Bilateral renal cortical necrosis

Table 5. *Pathologic changes in "certain virus infections", complicated by severe haemorrhage.* (McKay and Margetten, 1967)

Brain: 1. No changes
      2. Capillary platelet thrombi
      3. Capillary fibrin thrombi
      4. Perivascular ring haemorrhage
      5. Focal infarction

Pituitary:
      1. No changes
      2. Capillary fibrin thrombi
      3. Focal haemorrhage
      4. Extensive infarct necrosis

Adrenals:
      1. No changes
      2. Platelet or fibrin thrombi in the sinusoids
      3. Focal haemorrhage or necrosis
      4. Diffuse haemorrhagic necrosis

Gastro-intestinal tract:
      (The mucosa is predominantly involved)
      1. No changes
      2. Petechiae or echymoses
      3. Small focal ulcers
      4. Large multiple ulcers or pseudo-membranous enterocolitis

---

develop which are to a certain degree phagocytized by the RES, particularly by the Kupffer cells. If the surface of the platelets is damaged by adsorption of viruses, they tend to aggregate, as can be shown *in vitro* in the case of myxoviruses. In the case of endothelial lesions, adhesion, and clumping of thrombocytes may also take place.

Thus a pathological viscous metamorphosis is induced, which either liberates thromboplastic material or diminishes retraction capacity. This process can contribute to the occurrence of disseminated intravascular coagulation with its sequelae.

The coagulation defects demonstrated in Table 6 suggest intravascular clotting.

Table 6. *Criteria for intravascular coagulation*

A. Decrease of plasmatic clotting factors.
   Factor VIII (with prolongation of PTT and
                         recalcification-time)
   Factor V
   Factor II   (Prothrombin)
   Factor VII and X

B. Thrombocytopenia
   Loss of ATP
   Loss of Serotonin
   Loss of platelet factor 3

C. Decrease in Antithrombin III

D. Secondary Fibrinolysis with increase of
   Fibrinogen degradation products (Anti-
   thrombin VI)
   Prolongation of Thrombin time

E. Phagocytosis of Fibrinogen degradation products in the RES

Using coagulation methods in guinea pigs, which were previously infected with the Marburg virus, we tried to find out additionally plasmatic coagulation defects. The guinea pigs became ill on the first or second day, developing a fever which usually persisted for four to five days. In the case of high passage numbers, all the animals died on the fifth or sixth day after the onset of pyrexia. Short before death the plasma of the animals became strongly lipaemic and the native blood did not coagulate any more, as has been reported by SMITH et al. Coagulation was prolonged, however, if calcium ions were added to the native blood.

As is shown in Table 7, a reduction in a number of plasmatic coagulation factors took place during the course of the illness: on the third day of fever the level of factor VIII fell sharply, as well as the number of thrombocytes and the levels of factors II, V, VII and X. Mixing experiments were carried out to exclude the presence of a factor VIII inhibitory substance which yielded negative results.

Table 7. *Coagulation-studies in guinea pigs infected with Marburg virus*

Assay of clotting factors guinea pig no.	Before infection	1. day	2. day			3. day		4. day			5. day			6. day with fever
		151	176	171	252	150	144	151	174	141	151	148	188	176
Factor VIII	100—60% = 39—34 sec	70—100%	62%	34%	100%	24%	5%	18%	9%	8%	7%	6%	13%	7%
PTT	25—35	25	30	33	30	44	39	43	59	58	118	58	114	172
Recalcification time (sec)	40—80	67	83	67	47	109	84	93		116	260	148	170	211
		normal	normal			pathologic		pathologic			pathologic			pathologic
Factor V	100—60%	44%	52%	28%	15%	1½%	1%	9%	9%	5%	2%	1%	1%	1%
Prothrombin	100—60%	60%	40%	29%	—	5%	3%	27%	11%	18%	15%	2.5%	4%	19%
Factor VII	100—60%	40%	45%	52%	50%	17%	20%	18%	4%	10%	6%	6%	3%	5%
Factor X	100—60%		70%	48%	70%	20%	40%	18%	8%	13%	8%	5%	5%	8%
Thromboplastin time (sec)	36 44 33	46	58	77	41	86	89	88	228		175	100	150	185

Corresponding with the coagulation defects, the recalcification time, the partial thromboplastin time, and the thrombin time all showed pathological values. The third test always showed the formation of a normal thick clot which, however, had an unusual glassy appearance. In only two of the guinea pigs was the plasma fibrinogen reduced of 100—150 mg %. In the other cases the results remained within the normal range of 450—600 mg %.

It has to be mentioned, however, that the washing of the infectious and hyperlipaemic clots provided some difficulty.

In two of the guinea pigs we estimated the level of factor XIII before death by measuring the transglutamination reaction according to Loewy. We found it to be normal.

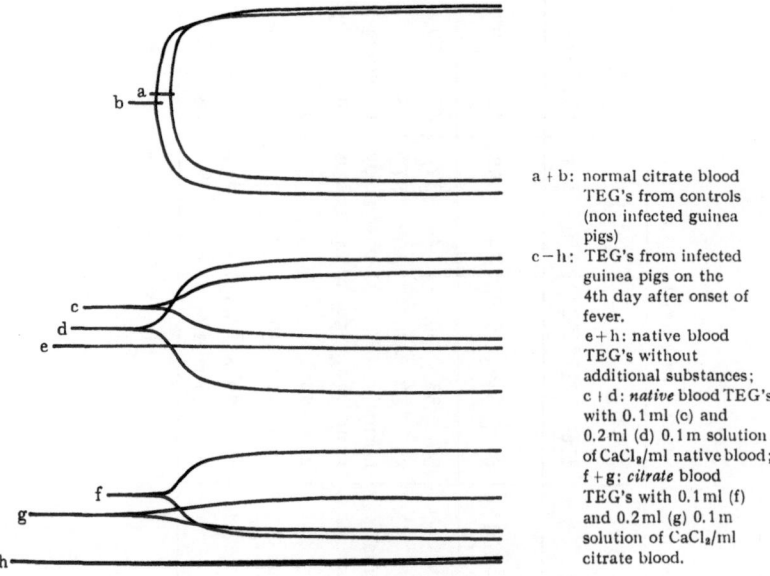

a + b: normal citrate blood TEG's from controls (non infected guinea pigs)

c — h: TEG's from infected guinea pigs on the 4th day after onset of fever.
e + h: native blood TEG's without additional substances;
c + d: *native* blood TEG's with 0.1 ml (c) and 0.2 ml (d) 0.1 m solution of CaCl₂/ml native blood;
f + g: *citrate* blood TEG's with 0.1 ml (f) and 0.2 ml (g) 0.1 m solution of CaCl₂/ml citrate blood.

Fig 1. Thrombo-elastograms of one normal guinea pig (top) and two guinea pigs infected by Marburg Virus

The pathological curves (Fig. 1) show a prolongation of the reaction time (r-time/ TEG), of the clot formation time (k-time/TEG), and a lowering of the maximal amplitude (m_ε/TEG). While TEG's of native blood showed no evidence of coagulation, on the 4th day after the onset of fever coagulation did occur, when calcium was added, depending on its concentration (Fig. 1 c + d). A simular result was obtained on testing citrated blood (Fig. 1 f + g). In spite of the marked thrombocytopenia, the maximum amplitude in the TEG increased during a prolonged course. Generally this result is not obtained with thrombocytopenia. It may be due to the different nature of the guinea pig fibrin or its stabilization and elasticity. There was no evidence in the TEG of spontaneous fibrinolysis, nor could spontaneous lysis of recalcified clots be observed.

Table 8.   Thrombin time in guinea pigs with Marburg virus disease done with different concentrations of thrombin

	Before infection	1. day	2. day	3. day	4. day	5. day	6. day with fever
guinea pig no.		151	176 171 252	150 144	151 174	151 148 188	176
Concentration of Thrombin							
1.4 NIH units	12.0 15.5 13.0	17.1	12.6 16.8 11.8	16.0 17.5	16.8 24.0	29.1 19.0 24.6	26.0
0.88 NIH units	19.8 21.8 19.0	21.0	25.0 27.2 19.0	18.3 23.5	22.5 40.0	40.5 28.4 50.4	Ø
0.6 NIH units	34.5 41.2 32.1	40.0	30.2 42.0 31.0	51.0 64.8	51.8	Ø 28.4 Ø	Ø

Ø = no coagulation

Table 9.   Platelet count and clot retraction in guinea pigs with Marburg virus disease

		1. day	2. day	3. day	4. day	5. day	6. day with fever
guinea pig no.		151 171	176 171	150 144	151 141	151 148 188	176 189
	Normal						
Platelet count 350 — 500 × 10³		412 367	213 198 154	83 118	125 109	77 25 86	53 69
Clot retraction (cm)	5.0 — 6.5	5.0 5.7	5.2 5.3 4.2	1.5 3.0	3.0 1.5	0.5 2.0 1.5	2.5 0.5

On heated fibrin agar plates, however, the lysis halos were larger when a euglobulin fraction of an infected animal was tested and thus indicated an increased fibrinolysis.

As is shown in Table 8, we got an indirect confirmation that fibrinogenolysis took place when we determined the thrombin times with different thrombin concentrations. The thrombin times became longer after the third day of fever. Now precipitation of fibrin could be achieved when using low concentration of thrombin (0.6 NIH units) in three guinea pigs (151, 188, and 176). The prolongation of the thrombin times can only be caused by the formation of antithrombin VI, which occurs during an increased secondary fibrinolytic activity in the blood. Finally, the number of platelets and the retraction are compared. Table 9 shows that from the third day of fever a severe deficiency of retraction occurs which does not quite correspond with the decrease in the numbers of platelets.

The pathological mechanisms of coagulopathy should be clarified by further investigations, including studies of calcium balance, determinations of blood fats, histological demonstration of fibrin in blood vessels of various organs, estimations of fibrinogen and factor XIII. Then it may be possible to explain the severe bleeding tendency in the lethal cases.

In conclusion

1. The severe coagulopathy and thrombocytopenia which occur in guinea pigs after infection with the Marburg virus have to be regarded as a disseminated intravascular coagulation. This is proved by the platelets and fibrin thrombi in the affected organs which show the same necrosis of the parenchymal cells as in human subjects.

2. These thrombi were demonstrated in three out of five patients who died (Gedigk et al., 1968).

3. In the cases with less pronounced thrombocytopenia, the plasma clots showed a severe disorder of retraction which was possibly due to damaged thrombocytes.

4. The pronounced hyperlipidaemia appears to be due to a blockade of the RES and its specific function. This is often observed in cases of consumption coagulopathy.

5. The blood of dying guinea pigs, does not coagulate shortly before death. This observation had already been made by Smith et al. It is, however, interesting that, after the addition of calcium ions in adequate concentration, a delayed coagulation occurs which is related to the decrease in the level of the coagulation factor. Whether these findings are related to the calcium deposits in several organs which were reported by Korb in 1969, has still to be clarified.

6. The coagulation analytical tests showed a marked reduction of nearly all coagulation factors.

References

DVILANSKY, A., BRITTEN, A. F. H., LOEWY, A. G.: Factor XIII Assay by an Isotopic Method. Brit. J. Haematol. **18,** 399 (1970).

GAGEL, CH., LINDER, M., MÜLLER-BERGHAUS, G., LASCH, H. G.: Virus Infection and Blood Coagulation. Thromb. diath. haemorrh. Vol. XXIII, 1—11 (1970).

GEDIGK, P., BECHTELSHEIMER, H., KORB, G.: Die pathologische Anatomie der Marburg-Virus-Krankheit. Dtsch. med. Wschr. **93,** 390 (1968).

JERUSHALMY, Z., KOHN, A., DE VRIES, A.: Interaction of Myxoviruses with Blood Platelets in vitro. Proc. Soc. exp. Biol. **106,** 462 (1961).

McKAY, D. G., MARGETTEN, W.: Disseminated Intravascular Coagulation in Virus Diseases. Arch. intern. Med. **120,** 129 (1967).

KORB, G., and SLENCZKA: Histological Findings in Liver and Spleen of Guinea-Pigs after Infection by the Marburg-Virus. Marburg Virus Disease. Berlin/Heidelberg/New York: Springer 1970.

LINDER, M., GAGEL, CH.: Virusinfektion und Blutgerinnung. Thrombos. Diath. haemorrh. XX, **3/4,** 603 (1968).

LOEWY, A. G.: Personal Communication.

MARTINI, G. A., KNAUFF, H. G., SCHMIDT, H. A., MAYER, G., BALTZER, G.: Über eine bisher unbekannte, von Affen eingeschleppte Infektionskrankheit: Marburg-Virus-Krankheit. Dtsch. med. Wschr. **93,** 559 (1968).

SMITH, C. E. G., SIMPSON, D. J. H., BOWEN, E. T. W., SLOTNIK, J.: Fatal Human Disease from Vervet Monkeys. Lancet **1967,** II, 1119.

Pathologic Anatomy of the Marburg Virus Disease

P. GEDIGK, H. BECHTELSHEIMER, and G. KORB

In August and September, 1967, in Marburg, Frankfurt, and Belgrade, there appeared in institutes experimenting with animals serious human illnesses caused by contact with monkeys. Twenty-seven people were affected. Only in Marburg 23 persons were taken ill and, of these, 20 had had direct contact with blood, organs, or cell-derived cultures from monkeys imported from Uganda, of the type known as Cercopithecus aethiops. Three patients were infected through contact with people already sick. The period of incubation, which could be accurately determined in patients who had had only a single contact with infectious material, was of five to seven days' duration.

Twenty-five patients recovered slowly after being ill for approximately 15 days. Five patients died between the 8th and 16th day of illness.

In the autopsies of the five patients who died in Marburg, we attempted to assemble the morphological findings. In order to deduce the development of the morphological alterations, we arranged the cases according to the duration of sickness, i.e., according to the elapsed time between the appearance of the first clinical symptoms and the moment of death. We then compared and correlated the results of these autopsies.

In all of the deceased, it was possible to ascertain macroscopically a pronounced hyperemia of the leptomeninges and an edematous brain swelling, accompanying the usual signs of central death. In addition, signs of a hemorrhagic diathesis were present in the form of considerable bleedings into the skin and mucous membranes and occasionally also into the soft tissues and parenchymatous organs. With the exception of the last case, the stomach and large sections of the intestines were filled with blood and in part also with black stool, although it was impossible to discover any source of bleeding, such as ulcers or erosions.

In none of the patients was the liver atrophic. It had in all cases a decidedly firm consistency and macroscopic analysis revealed no pathological findings worth mentioning.

It should be stressed that in none of the cases there was an icteric or subicteric discoloration of the liver, of the skin or of other organs. The gallblader was noticeably enlarged and tightly filled. Hence the autopsies revealed, in agreement with the clinical findings, no indications of significant disturbances of the bilirubin excretion.

The spleen was slightly enlarged in only two cases; the red pulpa was more frequently hardened, but revealed no irregularities macroscopically. In all cases, however, there was moderate swelling of the lymph nodes, which was especially evident on the hili and in the abdomen. The kidneys showed pale swelling in all cases. Finally, all of the deceased had a dark, blue-reddish, livid discoloration in the region of the external genitals near the scrotum or vulva.

The morphological picture of the lungs was not uniform. In general, the parenchyma of the lungs appeared dry. In two cases, there was an unevenly developed bronchopneumonia. One case revealed scars and areas of fibrosis as well as bronchiectasis and an emphysema which had certainly originated long before this viral disease. The hearts of all the deceased were dilated.

On surveying the macroscopical findings, it is evident that these morphological changes, taken separately or in combination, did not show a characteristic pattern which would allow classification in a well-known group of diseases.

The *histological examination* of the organs provided considerably more information. Surveying the histological findings, we are able to define six groups of alterations:

1. In almost all organs—with the exception of the skeletal muscles, the lungs and the skeleton—focal *necroses* appeared, which as a rule were not accompanied by an inflammatory reaction. These necroses were more conspicuous in the liver (see Marburg Virus Hepatitis) and in the lymphatic system. We were not, however, able to determine immediately where these necroses first appeared. Moreover, there were well-developed necroses in the testicles and ovaries, whereas they reached only minimal size in the kidneys, the adenohypophysis, the thyroid gland, the suprarenal gland and the skin. In all organs, the parenchymal cells were affected more by the necrobiosis than the mesenchymal structures.

The findings in the patients who died in the later stages of the illness and the results of numerous liver biopsies on convalescent patients allowed us to conclude that the necrosis—at least in the liver—could be relatively quickly replaced through regeneration in spite of their great extent.

2. The *lymphatic system* reacted in a special way. In addition to the aforementioned necrosis of follicles, the red pulpa of the spleen and the medullary part of the lymph nodes revealed a loss of cells and an embedment of a finely granulated eosinophilic material the nature of which has not yet been indisputably determined. We believe it possible that it contains some thrombocytic aggregates. There appeared later an increasing infiltration of plasmacellular elements and to some extent also of monocytes. The same cells were also discovered in large numbers in the mucous membranes of the stomach and intestines. It seems probable that this plasmacellular and monocytoidal infiltration occurred in connection with an immunological process.

3. A constant finding in all autopsied cases was the appearance of peculiar *basophilic bodies*. They were predominantly round in form and had, as a rule, a diameter of $1-2$, at most $3-4$ μ. Repeatedly, one had the impression that the basophilia covered only a crescent-shaped area, surrounding an eosinophilic core. The basophilic bodies were found especially within the area or in the vicinity of necroses, in cell phagocytes but also extracellularly.

Occasionally, they could also be found independently from necroses in the cytoplasma of otherwise apparently intact liver cells and kidney epithelia and also in the immediate vicinity of capillaries. Their coloration and histochemical properties indicated a high nucleic acid content, as it appears to the same degree in cell nuclei. These characteristics led at first to the conclusion that we were dealing with free or phagocytic fragments of necrotic cells; on second thought, however, we suppose that so-called inclusion bodies were present, as they have often been observed in many viral diseases. This question remains to be investigated by further histochemical reactions and electron-microscopical studies.

4. Furthermore, it was without exception possible to determine serious parenchymal damage of the kidneys, accompanied by signs of a tubular insufficiency.

5. In agreement with the macroscopical findings, we discovered histological signs of a hemorrhagic diathesis.

6. Histological sections of the brain revealed a diffuse encephalitis with a perivascular lymphocytic infiltrate. — In one case, there was a massive hemorrhage of the corpus callosum.

Remarkable was the minimal or almost lacking damage to the heart muscle. The interstitial edema, which we discovered histologically in all cases, was as a rule not accompanied by significant damage to the heart muscle. We would like to consider this edema as the result of an increased capillary permeability, but not as a sign of myocarditis. We must admit, however, that such an edema can clinically very closely resemble the signs of a myocarditis. The macroscopically determined weakening and dilatation of the heart ventricles is doubtlessly the consequence of the terminal failure of the heart and circulatory system.

Next to the determination of the decisive disease and of possible complications, there arises above all the question of the actual cause of death. In the case of some of the patients who died from the Marburg virus infection, we must suppose multiple causes for death.

The essential elements of the Marburg virus disease have been characterized in the preceding paragraphs:

1. Extensive necroses, 2. plasmacellular, monocytoidal transformation of the lymphatic tissue, 3. ubiquitous, basophilic bodies, 4. tubular kidney-insufficiency, 5. hemorrhagic diathesis, 6. cerebral damage.

These elements were developed to *different* degrees in all the deceased patients, but in their basic characteristics, they were uniformly present. They revealed only individual variations, such as are usually found in every illness due to the varying defensive reaction of the patients, but also to the duration of the illness and possible secondary diseases.

In all of the deceased the cerebral damage doubtlessly played an essential role. Moreover, we have to stress the serious intoxication or general metabolic disturbance, in part due to extensive parenchymal necrosis and their detritus and in part to liver damage or tubular insufficiency of the kidneys. It is propable that the general intoxication was also an important contributing factor in the cerebral damage and also responsible for the failure of the heart and circulatory system. On the basis of the anatomical findings, it is not possible to decide what significance should be given to the quite pronounced hemorrhagic diathesis and the clinically observed diarrhoea.

In one case, the extensive purulent and hemorrhagic bronchopneumonia, a complication of the main disease, led to a great strain on the heart and must thus be considered as an important contributing cause of death.

Let us now return to the question put at the outset, i.e., whether or not the virus illness transmitted by the monkeys can be considered as a special, individual disease. If we now survey the presented findings in view of this question, it is clear that from the pathological-anatomical standpoint, no specific morphological changes could be found which do not appear in one form or another in other illnesses. Nevertheless, the overall picture was characteristic neither of yellow fever nor of any other well-known virus disease nor for a leptospirosis or rickettsiosis. Each of these diseases has some symptoms in common with the disease that we have been studying; however, they never coincide—completely—with the clinical and morphological picture of our illness. The extent and, above all, the combination and pattern of the alterations in the different organs were certainly the special and characteristic feature, and this justifies our speaking of a new and up to now unknown disease.

References

1. BECHTELSHEIMER, H.: Die Pathologische Anatomie der Marburg-Virus-Krankheit. Habilitationsschrift, Marburg 1968.
2. BECHTELSHEIMER, H., JACOB, H., SOLCHER, H.: Zur Neuropathologie der durch grüne Meerkatzen (Cercopithecus aethiops) übertragenen Infektionskrankheit in Marburg. Dtsch. Med. Wschr. **93,** 602—604 (1968).
3. BECHTELSHEIMER, H., JACOB, H., SOLCHER, H.: The Neuropathology of an Infectious Disease transmitted by African Green Monkeys (Cercopithecus aethiops). Germ. Med. Monthly, **14,** 10—12 (1969).
4. BECHTELSHEIMER, H., KORB, G., GEDIGK, P.: Die Marburg-Virus-Hepatitis (Untersuchungen bei Menschen und Meerschweinchen), Virchows Archiv, Abtlg. A, im Druck.
5. GEDIGK, P., BECHTELSHEIMER, H., KORB, G.: Die pathologische Anatomie der ,,Marburg-Virus''-Krankheit (sog. ,,Marburger Affenkrankheit''). Dtsch. Med. Wschr. **93,** 590—601 (1968).
6. GEDIGK, P., BECHTELSHEIMER, H., KORB, G.: The Morbid Anatomy of Marburg-Virus-Disease. Germ. Med. Monthly, **14,** 68—77 (1969).
7. GEDIGK, P., KORB, G., BECHTELSHEIMER, H.: Die Pathologische Anatomie der Marburg-Virus-Krankheit. Verh. Dtsch. Ges. Path., **52,** 317—320 (1968).
8. KORB, G., BECHTELSHEIMER, H., GEDIGK, P.: Histologische Befunde bei der ,,Marburg-Virus''-Krankheit (sog. ,,Marburger Affenkrankheit''). Dtsch. Ärztebl. **65,** 1089—1096 (1968).
9. KORB, G., SLENSKA, W., BECHTELSHEIMER, H., GEDIGK, P.: Tierexperimentelle Untersuchungen über die Entstehung und den Ablauf der Marburg-Virus-Hepatitis. Virchows Archiv, Abtlg. A, im Druck.

The Neuropathology of the Marburg Disease in Man

H. Jacob

With 9 Figures

   The clinical signs and symptoms of what has become known as the Marburg or green-monkey disease have been described in detail by Martini et al. (1968), and prominent among them are disturbances of consciousness, either transient or progressing to coma, which point to an involvement of the nervous system. This

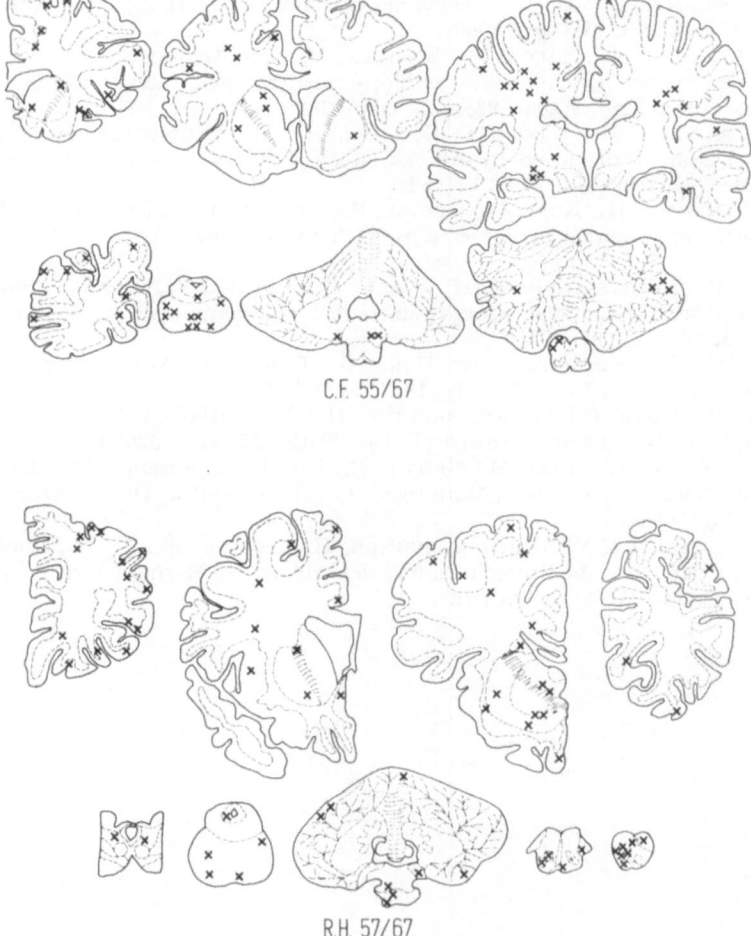

Fig. 1. Location of glial nodules. (Cases C.F. 55/67 and R. H. 57/67)

has been confirmed from whole-brain sections of the fatal cases C. F. (55/67—2225/67), R. H. (57/67—2188/67) and H.-O. V. (55/67—2189/67), sections which include the cerebral and cerebellar hemispheres, the basal ganglia and the brain stem down to the medulla oblongata. These brains had the typical appearance of a glial-nodule encephalitis, distributed throughout the brain, and combined with

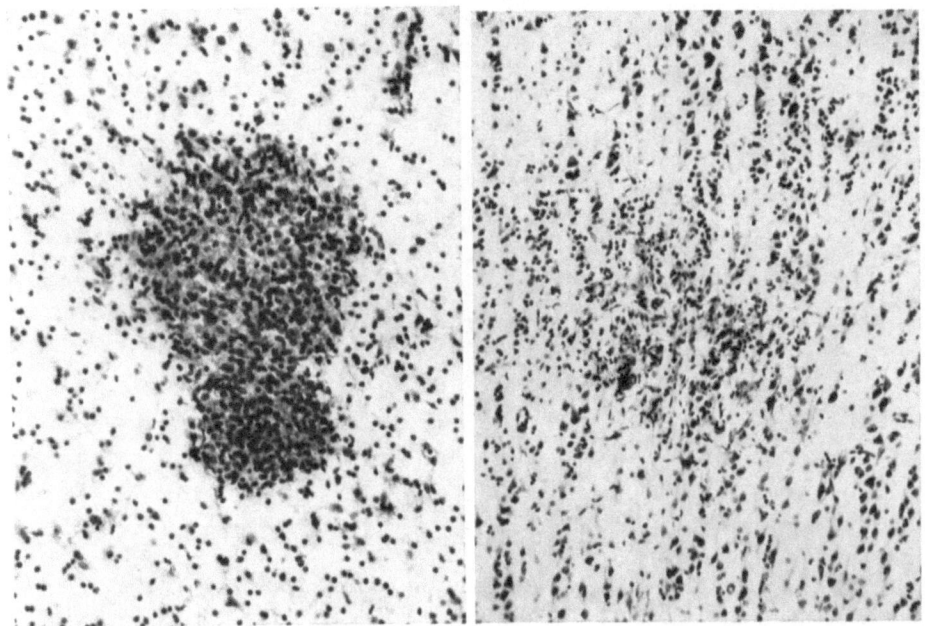

Fig. 2. Dense, richly cellular, relatively sharply defined glial-nodule in the white substance. (C.F. 55/67)

Fig. 3. Syncytial focus of glial cells with dispersed rod cells in the neighbourhood. (C.F. 55/67)

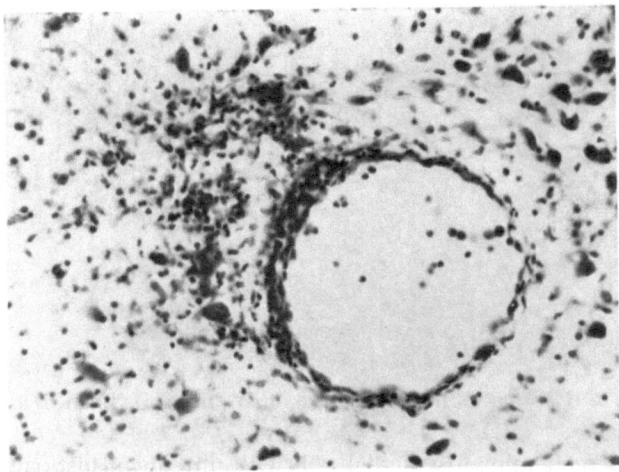

Fig. 4. Dense glial proliferation with slight lymphocytic perivasal infiltration in the thalamus. (C.F. 55/67)

a diffuse alteration and proliferation of the glial cells, as well as a relatively discrete perivascular lymphocytic inflammation. The glial nodules were commonly but not invariably associated with vessels showing lymphocytic infiltration, and they

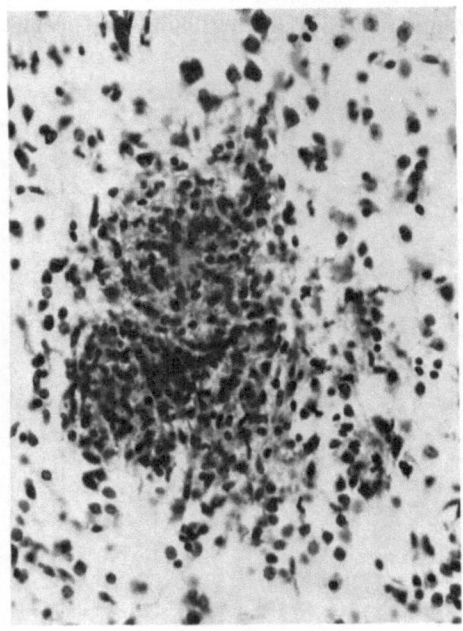

Fig. 5. Dense glial nodule in the cortex. (C.F. 55/67)

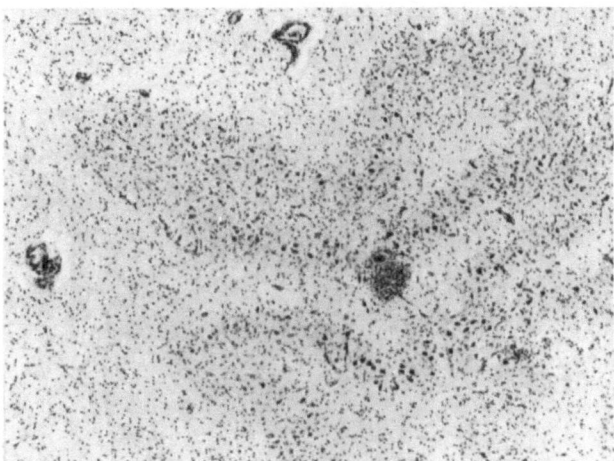

Fig. 6. Dense glial nodules in the oliva inferior. (R.H. 57/67)

were composed predominantly of glial cells with infrequent histiocytic-epitheloid elements, and they varied in size from a few cells up to large collections of glial cells as large as 300 μ in diameter.

They can be divided into three types on the basis of their structure and cellular composition: 1. dense, richly cellular, relatively sharply defined nodules composed of astrocytes with large, round to oval nuclei, some partly sausage-shaped, with

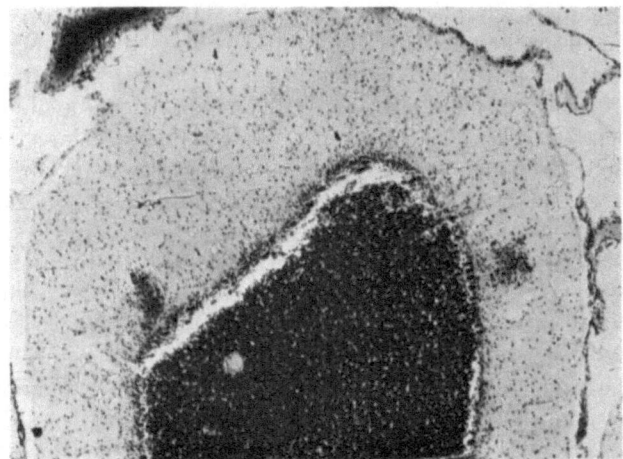

Fig. 7. Two dense glial nodules in the molecular layer of cerebellar cortex. (C.F. 55/67)

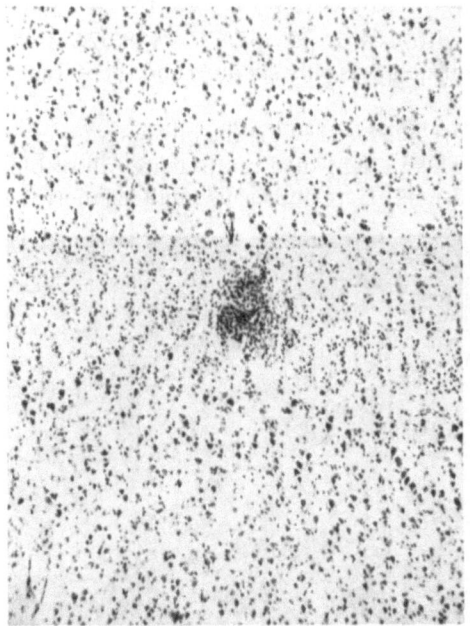

Fig. 8. Dense glial nodule in lamina III/IV of the cerebral cortex. (C.F. 55/67)

sparse chromatin, and less frequent oligodendrocytes with nuclei rich in chromatin, and isolated histiocytic elements; Fig. 2, 5, 6, 7, 8; 2. looser ill-defined "syncytial" foci and glial cells, in which microglia in the form of rod cells together with astrocytes and oligodendrocytes are proliferating: the rod cells are dispersed

in the neighbourhood around the edges; Fig. 3; and 3. collections initially of glial
cells with a denser or looser arrangement but never of the nature of "glial stars".

Inside the glial nodules pyknotic and karyorhexic changes are seen only in
the nuclei of oligodendrocytes with many nuclear fragments. Fig. 2 and 5.

In general, no necrotic appearances are seen. The cytoplasm of the glial cells
is ill-defined and either elongated or round, and the cells lie thickly upon one another.
The myelin sheaths in the white matter are, like the axons, separated from one
another, and often shrunken but not destroyed, and no definite breakdown products
are found associated with them. Neurones inside the nodules are well preserved.

Even inside diffuse proliferations of glia, it is the astrocytes and oligodendrocytes which predominate. Sometimes microglia and sometimes proliferations of
rod cells can be seen.

The extension of the process in the grey and white matter affects the whole
central nervous system down to the medulla oblongata in the form of a panencephalitis, also the emerging cranial nerves. A particularly heavily involved region
in two cases was the pons. It is noteworthy that there were localized aggregations
of nodules in the cerebellar cortex as well as in the cerebral cortex and white matter.
Fig. 2 gives an approximate general picture of the brains in cases C. F. (55/67 —
2225/67) and case R. H. (57/67 — 2188/67). It was possible to make meaningful
comparisons on a clinical and neuropathological basis between case C. F. (55/67 —
2225/67), with a ten-day course of neurological symptoms — a kind of "encéphalite
comateuse" (Rimbaud, Passonant et Vallat (1951)) — and a widely distributed
glial-nodule process, with case R. H. (57/67 — 2188/67) in which the duration of
disturbance of consciousness was only six days and there were undoubtedly far less
marked changes.

In the case H.-O. V. (56/67 — 2189/67) the process was very sparse and obviously
in the initial stage. This patient died of an acute fall of blood pressure without
neurological manifestations. Here numerous small haemorrhages in the corpus cal-

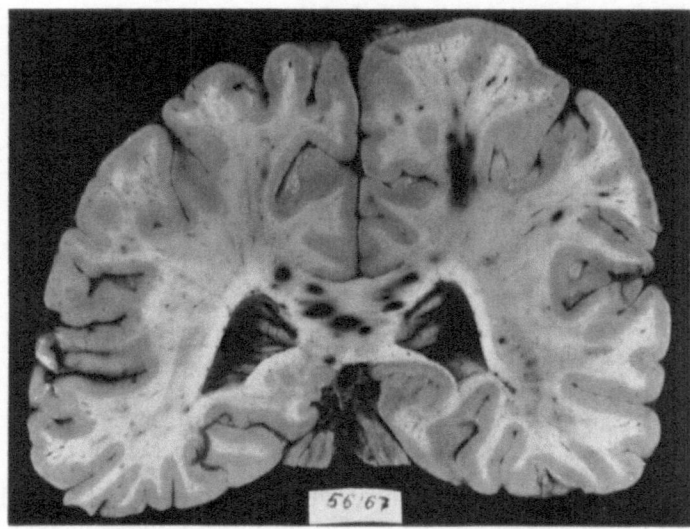

Fig. 9. Multiple haemorrhages in the splenium corporis callosi and in the white substance of parieto-occipital lobe. (V. 56/67)

losum and cerebral white matter constituted the neuropathological picture. Fig. 9.
A possible approach to this acute massive haemorrhagic appearance is provided by
two samples of cerebral cortex and white matter taken at necropsy from cases
K. F. (377—2217/67) and J. R. (378—2218/67). Both patients had died after only
one day of rapidly deepening coma, and the brain samples showed neither glial
nodules nor any progressive changes in the glia. On the other hand, there was a
marked vascular congestion with isolated serous effusion and diapedesis of erythrocytes around the thin-walled and dilated blood vessels. In the two first
cases, with the longer clinical course, such leakages were only isolated ones, whereas
the general haemorrhagic diathesis in cases H.-O. V. (56/67—2189/67), J. R.
(378—2218/67), and K. F. (377—2217/67) sufficed to explain the appearance of
cerebral haemorrhage. The clinical course was comparable with the "syndrome
malin" described by ALAJOANINE et al. (1938) in postinfectious haemorrhagic
diathesis. By way of contrast, the other two cases had a course essentially similar to
that typical of subacute comatose encephalitis.

In terms of comparative neuropathology the encephalitic process of the "Vervet-
monkey disease" in man is reminiscent of a whole series of glial-nodule encephalitides caused by different agents. In the first instance, it resembles the picture of
typhus encephalitis HALLERVORDEN (1943) but differs from it in the cellular
composition of the nodules, in its minor lymphocytic component, especially in
the meninges, and in the lack of glial infiltration and of destruction of Purkinje
cells, and in the sparing of the dentate nucleus. It is also reminiscent of the large
group of arbovirus glial-nodule encephalitides such as American Eastern and Western equine encephalitides (VAN BOGAERT (1958)), St. Louis encephalitis, Australian Murray Valley encephalitis (Australian 'X' disease) (HAYMAKER, LEWIS and
SCHWARZ (1957)) Japanese B encephalitis (JACOB (1956)), the Russian spring-
summer encephalitides, and central European tick-borne encephalitis (SEITEL-
BERGER und JELLINGER (1960/1966)). It is true that there are certain differences
which will be mentioned in detail in another publication (JACOB und SOLCHER
(1968)), but they may in general be recognized by the fact these forms of encephalitis have a much greater complexity of their overall picture, characterized by
intensive mesodermal inflammatory manifestations, by destruction of neurones
with neurophagia and focal necroses or spongiform destruction, and these features
differentiate them from the so-called "pure" glial-nodule encephalitides. One is
also reminded of several malignant illnesses in typhoid or paratyphoid fevers,
and of sporadic cases of glial-nodule encephalitides of uncertain aetiology, as have
been described by PETTE (1942) and VAN BOGAERT, RADERMECKER, HOZAY et
LOWENTHAL (1961) (cf. JACOB (1959/1961)).

With regard to animal transmission experiments, it should be emphasized
that a relatively pure glial-nodule encephalitis can be produced in monkeys by
the same virus which causes a quite different kind of encephalitis in man, as has
been shown, for example by HAYASHI (1935) in Encephalitis japonica B and by
SCHEID, JOCHHEIM and STAMMLER (1956) in lymphocytic choriomeningitis (see:
JACOB (1956, 1958, 1961)). For this reason one must be prepared for the possibility
of alterations in the neuropathological picture in the course of consecutive animal
passages (compare the findings of ZLOTNIK and SOLCHER in this review!). Our
findings suggest that, in fact, in the course of the infection there is a change in the

agent with a corresponding change in the reactions in the central nervous system; it remains to be decided whether the agent multiplies in the central nervous system and hence behaves as a neurotropic virus in the strict sense (DOERR).

Summary

A report is given on the neuropathological findings in lethal cases of a severe exanthematous infectious disease which occurred in August to September 1967 in Marburg (Germany) and which was transmitted to man by green long-tailed monkeys from Uganda *(Cercopithecus aethiops)*. Two lethal cases showed typical clinical signs of a subacute comatose encephalitis. The histological picture was that of a panencephalitis with glial nodules ("Gliaknötchenencephalitis") and slight perivasal lymphocytic infiltration. A similar but less marked glial process in combination with focal haemorrhages in the cerebral white matter and corpus callosum was found in a case without unconsciousness and neurological signs. Slight diapedeses were seen in cases which had been unconscious only for one day.

The author compares these neuropathological findings with other forms of glial nodule encephalitis, i.e. in typhoid fever, typhoid and paratyphoid diseases, numerous arthropod-borne encephalitides, and in the glial-nodule encephalitis occurring sporadically in Central Europe. The question of a possible change in the structure of the encephalitic process after animal passage is discussed.

References

ALAJOUANINE, TH., MARQUÉZY, R. A., HORNET, T., LADET, J.: Die Läsionen des Nervensystems im Verlaufe des Syndrome malin toxinfectieux in der Kindheit. Bull. Soc. Hop. Paris **54,** 1512 (1938).

BOGAERT, L. VAN: Les encéphalitides verno-estivales. — Encéphalites d'origine inconnue. In: Hdb. d. spez. path. Anat. und Histol. (LUBARSCH, HENKE, RÖSSLE), Bd. XIII, 2. Teil, S. 362−397. Berlin−Göttingen−Heidelberg: Springer 1958.

BOGAERT, L. VAN, RADERMECKER, J., HOZAY, J., LOWENTHAL, A.: Encephalitides. (Proc. of a symp. on Neuropathology, Electro encephalography and biochemistry of Encephalitides) Antwerpen: Elsevier Publ. Comp. 1961.

HALLERVORDEN, J.: Die pathologisch-anatomischen Veränderungen im ZNS beim Fleckfieber. Militärarzt **8,** 26 (1943).

HAYASHI, M.: Übertragung des Virus von Encephalitis epidemica japonica auf Affen. Folia psychiat. neurol. jap. **1,** 419−465 (1935).

HAYMAKER, W., LEWIS, K. L., SCHWARZ, E.: Pathology of the Meningoencephalomyelitides due to viruses, Rickettsiae, Trypanosomes, Toxoplasma and Fungi. III. Internat. Congress on Clinical Pathology, Belgium 1957.

JACOB, H.: Die postinfektiösen sekundären Encephalitiden und Encephalopathien. Fortschr. Neurol. Psychiat. **24,** 244−274 (1956).

JACOB, H.: Neuropathologie der Viruserkrankungen des Zentral-Nervensystems. Dtsch. Zschr. f. Nervenheilk. **182,** 472−491 (1961).

JACOB, H.: Sporadische, atypische „primäre" Encephalitiden. Encephalitis japonica B und parainfektiöse Encephalitiden. Folia psychiat. neurol. jap. **61,** 311−341 (1959).

LHERMITTE, F.: Les Leucoencephalites. Paris: Editions Méd. Flammarion 1950.

MARTINI, G. A., KNAUFF, H. G., SCHMIDT, H. F., MAYER, G., BALTZER, J.: Über eine bisher unbekannte, von Affen eingeschleppte Infektionskrankheit: Marburg-Virus-Krankheit. Dtsch. med. Wschr. **93,** 599−571 (1968).

Pette, H.: Die akut entzündlichen Krankheiten des Nervensystems. Leipzig: G. Thieme 1942.

Rimbaud, L., Passonant, P., Vallat, G.: Die komatöse Encephalitis. Rev. neurol. **84,** (1950). — Presse méd. 1951, 605.

Scheid, W., Jochheim, K. A., Stammler, A.: Tödlicher Verlauf einer Infektion mit dem Virus der lymphocytären Choriomeningitis. Dtsch. Z. Nervenheilk. **174,** 123—139 (1956).

Seitelberger, F., Jellinger, K.: Frühjahr-Sommer-Encephalomyelitis in Mitteleuropa. Nervenarzt **31,** 49—60 (1960).

Seitelberger, F., Jellinger, K.: Neuropathologie der Zeckenencephalitis. Neuropat. Polska **4,** 367—400 (1966).

Marburg Virus Hepatitis

H. Bechtelsheimer, G. Korb, and P. Gedigk

The pathological picture of Marburg-virus disease was described and discussed in the previous papers. Despite the obviously pantropic nature of the virus, damage to the liver was prevalent; therefore the histological features of the accompanying hepatitis are worth investigating. By comparing the findings in inoculated guinea pigs, the autopsies of the patients in whom the disease was fatal and the results of biopsies from convalescent patients, it was possible to define the formal genesis and development of the liver damage and the subsequent recovery. The earliest alterations caused by the liver infection are of special interest, because obviously in advanced cases they may be masked by additional and unspecific changes.

The first alterations were seen in twenty guinea pigs[1], which were killed between the 1st and 15th days after inoculation. The most marked liver damage was seen in two patients who died at the peak level of the SGOT. Signs of a decline in the disease were seen in three patients who died on the 14th to 16th days of their illness, four to seven days after the SGOT peak. The late changes could be observed in 15 convalescent patients on whom liver biopsies were performed between the 13th and 39th days of illness.

The examination of the guinea pigs revealed that the *earliest pathological changes* occurred two to three days after the inoculation. Aside from the activation of Kupffer cells in the parenchyma, only some acidophilic necroses of single cells were observed. The necrotic cells showed coagulation of the cytoplasm as well as pycnotic and disintegrated nuclei.

Sometimes nuclei were seen as faintly recognizable eosinophilic round particles. Sometimes liver cells were seen in which at first only part of the cytoplasm was coagulated—similar to the finding in yellow fever. At first, these necrotic cells were situated in the acinus structure; later on, they were eliminated into the sinusoids as so-called Councilman bodies. Sometimes they contained lipid droplets within their cytoplasm. The surviving liver cells showed a lower glycogen content and only a few fatty droplets. Within the sinusoids there were only some lymphoid and monocytoid cells as well as a few basophilic bodies.

The changes in the portal canals were minimal and could be neglected. They contained only a few lymphoid and monocytoid cells.

During the *further development of the liver damage*, the number of single-cell necroses and group necroses of different sizes increased. The group necroses developed like those in yellow fever, i.e. through concentric expansion from the single-cell necroses, the central cells showing the most advanced necrotic changes. Within the group necroses, necrotic Kupffer cells were sometimes recognized.

[1] We wish to thank Prof. Siegert, Dr. Shu, and Dr. Slenczka for the opportunity to make a histological examination of the guinea pig material.

During the course of the illness, the group necroses spread out and became larger. As they progressed toward the periphery, new single-cell necroses could be observed in the remaining parenchyma.

It is noteworthy that, in the cytoplasm of some liver cells, round and faintly eosinophilic and sometimes weakly basophilic PAS-positive and Feulgen-negative inclusion bodies occurred, both singly and in groups. They were situated mainly in the periphery of the cytoplasm and were separated from the rest of the cytoplasm by a small, clear halo.

Even at this stage, numerous mitoses could be observed in the undamaged liver cells. These mitoses were distributed in the acini partly without recognizable systematic formation, but partly also in the neighborhood of necrotic cells. At the same time, a striking percentage of supercontracted mitoses could be seen. In some of these cases the group necroses showed an infiltration of lymphocytoid cells, which also occurred in the sinusoids and in the portal tracts. In addition accumulations of small (1—2 mμ in diameter) basophilic particles of mostly round shape could be seen in varying amounts in the necrotic areas and in the portal canals.

The *greatest liver damage* was seen in two severely affected guinea pigs and in two deceased patients with extremely high transaminase levels. Here, both single-cell necroses and group necroses showed signs of disintegration. In the vicinity of these areas there were fresh single-cell necroses and intracellular hyaline condensation areas. In the parenchyma the fatty degeneration became more apparent. Mitosis and liver cells with two nuclei were seen more frequently. Bile casts were not observed; however, the canaliculi were sometimes filled up with a PAS-positive material, which was also stained by protein dyes—a phenomenon described by KÜHN as occurring in the early stages of hepatitis. The Kupffer cells contained the phagocytized products of necrotic parenchyma cells. Furthermore, the numbers of monocytoid and lymphocytoid cells and basophilic particles in the sinusoids and in the portal canals had increased. The veins in the portal canals contained an inflammatory infiltration within their walls.

In contrast to the human material, some of the liver cells of the guinea pigs contained a basophilic, amorphous material resembling calcium deposits, which was PAS-positive, Feulgen-negative, and orthochromatically stained by toluidine blue. Furthermore, these animals showed an unusual regeneration process consisting of the formation of tubular liver cells, resembling the so-called biliary hepatocytes.

In the three patients who died 14, 15, and 16 days after the onset of the illness, i.e. 4 to 7 days after the transaminase peak, *the further development of the necroses* could be studied. The single-cell necroses decreased and disappeared. The group necroses showed advanced disintegration. After the resorption of necrotic material, temporary hyperhemia ("Entlastungshyperämie") and collapse of the fibers was observed. The argentophile fibers remained intact. In the areas which had lost parenchyma, Kupffer cells were seen with abundant cytoplasm containing remnants of cells, lipopigment granules, and erythrocytes. Later on, the parenchyma defects were replaced by regenerating hepatocytes. Thus, regeneration was nearly complete seven days after the SGOT peak. The lymphocytoid infiltration of the portal canals disappeared slowly. The remaining parenchyma showed fine to medium-sized fatty droplets for some time, and slight siderosis as well as liver cells with two nuclei.

The stage of restitution of the liver damage could be observed in liver biopsies of convalescent patients. Comparing the histological alterations with the maximum transaminase levels during the illness, revealed remarkable correlations.

By this method it was shown that only slight and uncharacteristic alterations remained in the liver *when the maximum transaminase level had not exceeded 500 mU/ml*. Aside from a slight polymorphism and vacuolization of the hepatocytes, only a slight, patchy, fatty degeneration of the parenchyma remained from the earlier lesions. The remnants of the inflammatory reaction of the mesenchyma could be seen in some nodules of Kupffer cells as well as in a few lymphocytoid and plasma cells in the portal canals. The Kupffer cells occasionally contained some siderin or ceroid. Signs of regeneration were mitoses and hepatocytes with two nuclei.

When the transaminase levels had exceeded 500 mU/ml, four essential alterations were found:

1. increased infiltration of the portal canals by histiocytes and plasma cell elements;

2. pronounced fatty degeneration of the hepatocytes with medium-sized and large fatty droplets;

3. nodules of Kupffer cells;

4. a few single-cell necroses resembling Councilman bodies.

No proliferation of the bile ducts, fibrosis, or postnecrotic cirrhosis occurred.

Thus the morphological features of the healing process are obviously determined by the early and initial maximum damage to the liver occurring during the illness.

Surveying these results, it is evident that "Marburg-virus" hepatitis is distinguished by the following criteria: in the *parenchyma*, disseminated single-cell necroses and group necroses are prevalent, with simultaneous fatty degeneration of the hepatocytes. Further, cytoplasmic inclusions are typical features, mainly at the onset and during the peak of the illness. Quite striking are the numerous basophilic bodies, which are obviously fragments of nuclei, and the unusually numerous abnormal mitoses. In the intralobular *mesenchyma*, a notable postnecrotic proliferation of Kupffer cells is found, sometimes necrotic Kupffer cells could also be seen. In the portal canals there is a lymphocytic and monocytic infiltration. Noteworthy disturbances of bilirubin excretion are morphologically undetectable.

Comparison of Marburg-virus hepatitis with other types of hepatitis—by means of the criteria mentioned above—gives following picture: in the *parenchyma*, at first *single-cell* and later *group necroses* of the coagulation type are prevalent. It is remarkable that these necroses are disseminated in all parts of the acini without favoring any particular region. Many of these single-cell necroses have nearly the same features as the so-called Councilman bodies. Obviously, the single-cell necroses developed from focal cytoplasmic coagulation which later affected the whole cell. By radial expansion they subsequently conflowed into group necroses.

In type, distribution, and expansion, the initial single-cell necroses correspond to the initial pattern of nearly all viral necroses of the liver, e.g. yellow fever, some types of hemorrhagic fever, Kyasanur Forest disease, Korean fever, dengue fever, Bolivian fever, some cases of mononucleosis, and some types of animal hepatitis as well as human hepatitis.

As the liver damage increases, differences can be seen in the extent of the developing group necroses. However, the differences are obviously more of degree than of kind. In slight cases of Marburg-virus hepatitis, as in other types of hepatitis, the necrosis of the cells is limited to single-cell necroses. The group necroses which sometimes develop by radial expansion can also be observed in several other types of hepatitis, e.g. in yellow fever, where, however, they are situated mainly in the intermediate zones of the acini. Moreover, we always take into account that the initial features of the necrosis may be masked and modified by shock or other circulatory disturbances.

Quite another situation exists in human hepatitis. Here the necroses do not expand radially and do not develop into group necroses of the coagulation type. In the later stages of classical viral hepatitis, ballooning and lytic necroses of the hepatocytes are prevalent. Another difference is the absence of the so-called diffuse cell iron, which is characteristic of viral hepatitis. Only occasionally can a slight siderosis of the hepatocytes be observed.

Particularly striking is *the simultaneous fatty degeneration of the hepatocytes*, which may differ in intensity and persist longer than the parenchymal damage. It has not so far been decided whether this fatty degeneration of the hepatocytes represents only additional secondary toxic or nutritive damage or whether it is due to a disturbance of the cell metabolism directly associated with the virus infection. This hypothesis may be supported by the simultaneous occurrence of necrosis and fatty degeneration of the parenchyma in yellow fever, in the etiologically unelucidated Labrea hepatitis, in dengue fever, and Korean fever. On the other hand, even in severe and fulminant forms of human hepatitis, fatty degeneration of the hepatocytes never occurs.

We must stress the fact that the *cytoplasmic inclusion bodies* are characteristic of the pathological alterations of the hepatocytes. We are not certain whether they correspond to the areas of antigen which SLENSKA demonstrated by means of fluorescence microscopy. Furthermore, we are unable to decide whether or not they are specific for Marburg-virus hepatitis, because we have not had adequate experience with viral diseases outside the European area. As far as we can ascertain, such inclusion bodies have not been described in the literature in the so-called hemorrhagic fevers or yellow fever. They are not found in human hepatitis.

A remarkable phenomenon is the occurrence of many *basophilic bodies*. The particles are evidently the result of nucleoclasia. In the case of fragmention of liver nuclei, they are formed by the necrosis of hepatocytes; in the portal canals as well as in other organs they are obviously derived from lymphocytes.

It is well known that nucleoclasia also occurs in other virus infections, which cannot be discussed here. Nevertheless, this phenomenon is seen to such an extent in Marburg-virus hepatitis that we would define this unusually pronounced nucleoclasia as characteristic of this illness.

In addition, we found isolated particles of chromosome in supercontracted mitoses, as described by Altmann in liver cells after intoxication with colchicine. This phenomenon has been explained as a disturbance of mitosis which is expressed in an exploded prophase. Above all, the high percentage of abnormal mitoses is quite striking, so that it may be supposed that the large number of mitoses is caused not only by regeneration processes but perhaps also by direct stimulation by the virus.

Regarding the *mesenchymal reaction* in Marburg-virus hepatitis, it can be stated that the initial *Kupffer cell proliferation* described above may also occur in other types of viral hepatitis. But it is never as pronounced as in the early stages of human viral hepatitis. *The necrosis of the Kupffer cells*, which occasionally occurs in the area of the group necroses, can perhaps be explained as the result of a reaction accompanying the parenchymal damage; on the other hand, it could be interpreted as a direct result of viral necrosis. Thus, the observations of Miayi, Bang, Jones, Cohen and Smetana in cases of viral hepatitis in mice lead to the conclusion that necrosis of the Kupffer cells may precede damage to the liver parenchyma. In our cases, however, we were unable to observe these features.

The postnecrotic proliferation of the Kupffer cells which develops in the areas of resorbed group necrosis is quite striking. No such extensive proliferation of Kupffer cells is seen, either in human hepatitis or in other types of reactive hepatitis. Once again, it seems probable that we have to deal not only with a simple spodogenous reaction following necrosis of the parenchyma, but also with direct stimulation by the virus.

Concerning the type and cellular composition of the *inflammatory infiltrates in the portal canals*, it must be stressed that they are not specific for Marburg-virus hepatitis and can occur similarly in other viral diseases and in rickettsiosis. The only surprising feature is the discrepancy between the extent of the necrosis and the relatively slight infiltration of the portal canals. Furthermore, perivenous infiltration of lymphocytes and monocytic elements is well known in many viral diseases.

Another surprising finding in Marburg-virus disease is that, in spite of the severe parenchymal damage and the enormous rise in transaminase levels, there were *no noticeable abnormalities or disturbances of bile excretion*. Morphologically, no biliary necrosis or signs of cholestasis could be detected. As an indication of a very slight disturbance of bile secretion, PAS-positive material could occasionally be seen in the canaliculi; this is also seen in the initial stages of human viral hepatitis (Kühn). At first sight, it seems astonishing that such severe damage of the liver parenchyma is not followed by jaundice—but on the other hand it should be stressed that necrosis of the hepatocytes plays a far less important part in the development of jaundice than the diffuse alteration of the surviving liver cells. Alteration of the still living cells is essential to produce the clinical symptoms. Diffuse alteration of the hepatocytes seems less pronounced in Marburg-virus hepatitis than in yellow fever and in human viral hepatitis, as it can be seen also morphologically in the lack of ballooning of the cytoplasm.

In summary it can be stated that the pathological alterations found in Marburg-virus hepatitis are not specific and characteristic when examined individually and also occur in many other virus diseases of the liver. Nonetheless, *the morphological features of Marburg-virus hepatitis are specific in their overall picture, i.e. in their combination, and in the pattern and development of the liver lesions.* The pathology of Marburg-virus hepatitis is completely different from that of classical human hepatitis and yellow fever and—as far as our survey of the literature showed—also from the liver alterations in so-called hemorrhagic fever.

References

1. ALTMANN, H. W., HAUBRICH, J.: Über hepatozelluläre Mitosestörungen und Kerneinschlüsse nach wiederholten Colchicingaben. Beitr. Path. Anat. **81,** 355—394 (1966).
2. BANG, F. B., WARWICK: Mouse Macrophages as Host Cells for the Mouse Hepatitis Virus and the Genetic Basis of their Susceptibility. Proc. Nat. Acad. Sci. **46,** 1065—1075 (1960).
3. BECHTELSHEIMER, H.: Die Pathologische Anatomie der Marburg-Virus-Krankheit. Habilitationsschrift, Marburg 1968.
4. BECHTELSHEIMER, H., JACOB, H., SOLCHER, H.: Zur Neuropathologie der durch grüne Meerkatzen (Cercopithecus aethiops) übertragenen Infektionskrankheiten in Marburg. Dtsch. med. Wschr. **93,** 602—604 (1968).
5. BECHTELSHEIMER, H., JACOB, H., SOLCHER, H.: The Neuropathology of an Infectious Disease transmitted by African Green Monkeys (Cercopithecus aethiops). Germ. Med. Monthly **14,** 10—12 (1969).
6. BECHTELSHEIMER, H., KORB, G., GEDIGK, P.: Die Marburg-Virus-Hepatitis (Untersuchungen bei Menschen und Meerschweinchen). Virchows Archiv, Abtlg. A, im Druck.
7. GEDIGK, P., BECHTELSHEIMER, H., KORB, G.: Die Pathologische Anatomie der ,,Marburg-Virus''-Krankheit (sog. ,,Marburger Affenkrankheit''). Dtsch. med. Wschr. **93,** 590—601 (1968).
8. GEDIGK, P., BECHTELSHEIMER, H., KORB, G.: The Morbid Anatomy of Marburg-Virus-Disease. German Med. Monthly, **14,** 68—77 (1969).
9. GEDIGK, P., KORB, G., BECHTELSHEIMER, H.: Die Pathologische Anatomie der Marburg-Virus-Krankheit. Verhh. Dtsch. Ges. Path., **52,** 317—320 (1968).
10. JONES, W. A., COHEN, R. B.: The Effect of a Murine Hepatitis Virus on the Liver. Am. J. Path., **41,** 329—347 (1962).
11. KORB, G., BECHTELSHEIMER, H., GEDIGK, P.: Histologische Befunde bei der ,,Marburg-Virus''-Krankheit (sog. ,,Marburger Affenkrankheit''). Dtsch. Ärztebl., **65,** 1089—1096 (1968).
12. KORB, G., SLENCZKA, W., BECHTELSHEIMER, H., GEDIGK, P.: Tierexperimentelle Untersuchungen über die Entstehung und den Ablauf der Marburg-Virus-Hepatitis. Virchows Archiv, Abtlg. A, im Druck.
13. KÜHN, H. A.: Die formale Genese der Hepatitis epidemica nach Untersuchungen an Leberpunktaten. Beitr. Path. Anat. **109,** 589—649 (1947).
14. MIYAI, K., SLUSSNER, R., RUEBNER, B.: Viral Hepatitis in Mice. An Electronmicroscopic Study. Exper. Mol. Path., **2,** 464—480 (1963).
15. SLENCZKA, W., SHU, H.L., PIEPENBURG, G., SIEGERT, R.: Antigen-Nachweis des Marburg-Virus in den Organen infizierter Meerschweinchen durch Immunfloreszenz. Dtsch. med. Wschr., **93,** 612—616 (1968).
16. SMETANA, H. F.: The Histopathology of Experimental Yellow Fever. Virchows Archiv, **335,** 411—427 (1962).
17. SMETANA, H. F.: Pathologic Anatomy of Early Stages of Viral Hepatitis. In: Hepatitis Frontiers (Eds. HARTMANN, F. W. et al.), Boston, Little, Brown & Co. 1957.

Morphology, Development, and Classification of the Marburg Virus

D. Peters, G. Müller, and W. Slenczka

With 14 Figures

The transmission of the Marburg agent from monkey to man was governed by the occurrence of high infectivity in blood and organs of the animals [1, 2]. Rather high titers were also observed in infected guinea pigs [3, 4, 5]. It was the electron microscopic examination of blood and organs of guinea pigs and monkeys which finally led to the recognition and description of this new agent. On morphological criteria it was shown to be a virus with a highly complex structure related to the viruses of vesicular stomatitis and rabies [6, 7]. This finding has been confirmed by others [8—12].

Morphology

The electron microscopy was done on fixed guinea pig material prepared in Marburg, and on fixed preparations from monkeys made available by Professor R. Haas and Dr. G. Maass, Freiburg.

In sediments from guinea pig and monkey plasm which had been centrifuged at 20 000 g in a special centrifuge tube for particle counting [13] directly on to coated grids, negative staining revealed large amounts of virions (Fig. 1). Their shape varies from fairly straight strands with rounded ends to forms of a six or a horseshoe, or coiled structures exhibiting a roughly annular aspect. These circular particles prevail in animal blood and have an overall diameter of about 250 nm.

Two peaks were obtained from measurements of particles where the entire length was visible. The portion with lengths over 900 nm was about 10%. Values around the median of 665 nm were found at a markedly high frequency. The second peak with the double value of 1 300 nm proved through measurements of a large quantity of longer particles to be significant. Extreme lengths of up to 8 μm were observed. The diameter of extended viruses lay between 70 and 80 nm. The higher values reported earlier were incorrect because of shrinkage of the contrast medium at the border of the particles.

Particles not penetrated by the contrast medium appear light and do not exhibit internal structures. In virions, however, in which penetration has occurred, a complex organisation of high regularity becomes visible (Fig. 1). The main characteristics are a dark central axis of about 20 nm, a surrounding tubular element bearing a cross striation with a periodicity of about 53 Å and a layer of enclosing material. In the annular forms the tubular element is coiled up to produce circle- or spiral-like figures which are entirely enclosed as a circular unit by the envelope. The ends of the strands, although touching in the circular forms, apparently are not fused. Despite superimposition, at least one end can usually be located. Insofar as length measurements on the strands of the annular form are possible, values close to the normal length are obtained.

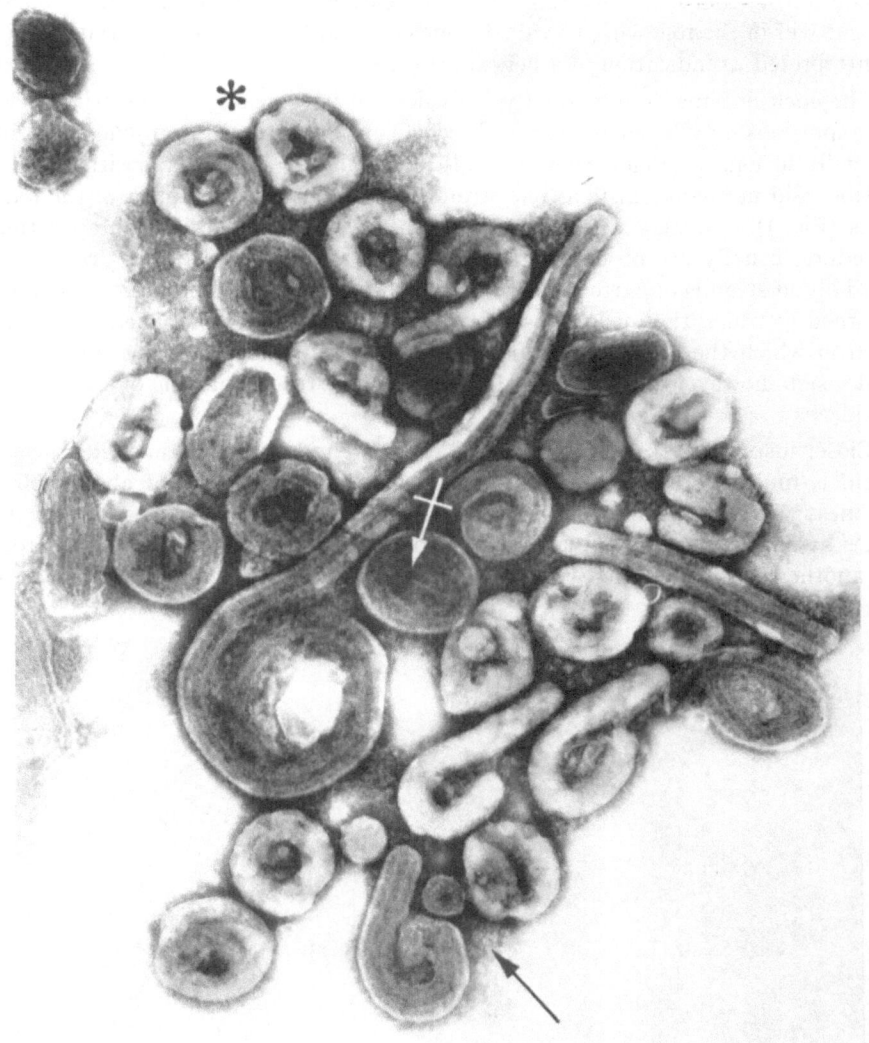

Fig. 1. Marburg virus, sedimented from guinea pig blood, 4 d p.i. Formaldehyde-fixed, negatively stained. Annular and 6-shaped particles besides one straight and one long form, with 2,7 μm 4-times longer than the normal virions. Axis and cross-striations visible where the contrast medium has penetrated. Spikes on the envelope surface (*), end-on view of a particle (→), transversing portion in a spiral form (↔).   60 000 ×

In the six-shaped forms the enclosing material surrounds the extended part closely, but the coiled portion is again most often enclosed in its entirety. In these particles disruption of the enclosing material may occur during preparation and as a rule this takes place at the site of the highest stress, the end of the coiled portion.

From such particles in which part of the tubular element is exposed, it can be deduced that this element has a diameter of about 30 nm. This value agrees well with measurements made on transversing portions in the spiral-like figures of the annular form (Fig. 1). Deviating from our earlier interpretation, this tubular element will in the following be called "nucleocapsid" and the cross striation will be interpreted as indication of a helical arrangement.

The enclosing material evidently is rather sensitive since particles with bleb-like expulsions may be encountered. The blebs mostly emerge from one end, but especially in long particles, they may also be found at any other position. The nucleocapsid normally exists as one strand. This is true also in most of the long forms (Fig. 1). Breaks, when they occur, perhaps induced by the preparation procedure, usually are observed in the coiled portion of the strand. However, especially in strands of extreme length, segments of the enclosing material can be discerned in which the nucleocapsid is missing. Very rarely branched forms are noted in which the enclosing material extends continuously in three directions. No decision, however, has been possible concerning a true branching of the nucleocapsid.

Closer inspection of the virion reveals that the material enclosing the nucleocapsid is multilayered (Fig. 2). The outer part is an envelope of about 100 Å thickness which stands out distinctly as a light zone. Its surface is studded with spikes having a size of about 70 Å arranged at a distance of roughly 100 Å from one another (Fig. 2a). These projections usually are best seen at the ends of the

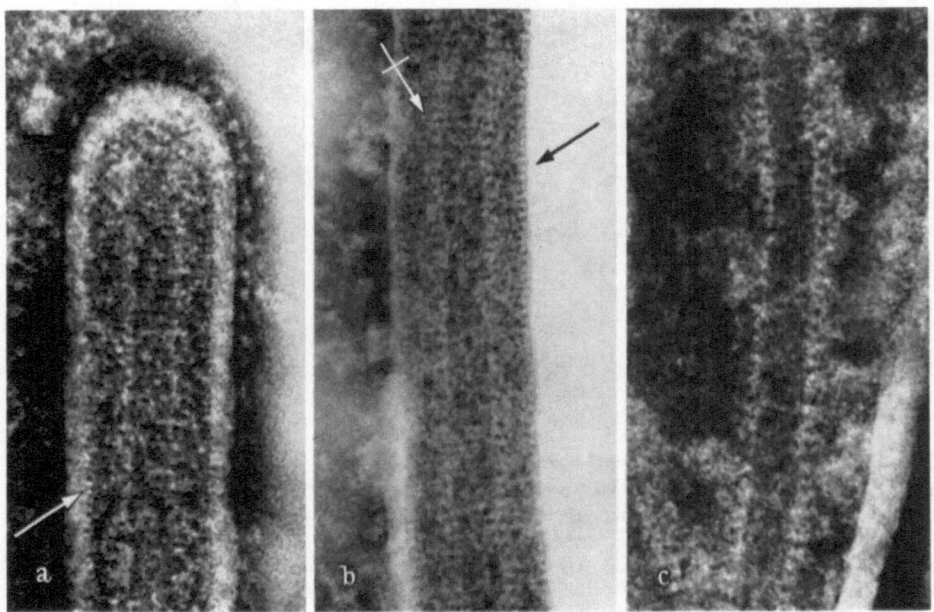

Fig. 2a—c. The same preparation as in Fig. 1. a) Envelope studded with spikes; cross-striation with 33 Å period (→); b) Particle segment without envelope; cross-striation with 53 Å period (↔); cross-striation with 33 Å period (→); c) Nucleocapsid without enclosing material. 350 000 ×

particles or on virions which are not penetrated by the contrast medium (Fig. 1). Below this envelope another cross-striation may be visible which has a significantly smaller periodicity in the order of about 33 Å (Fig. 2a). It stands out most clearly

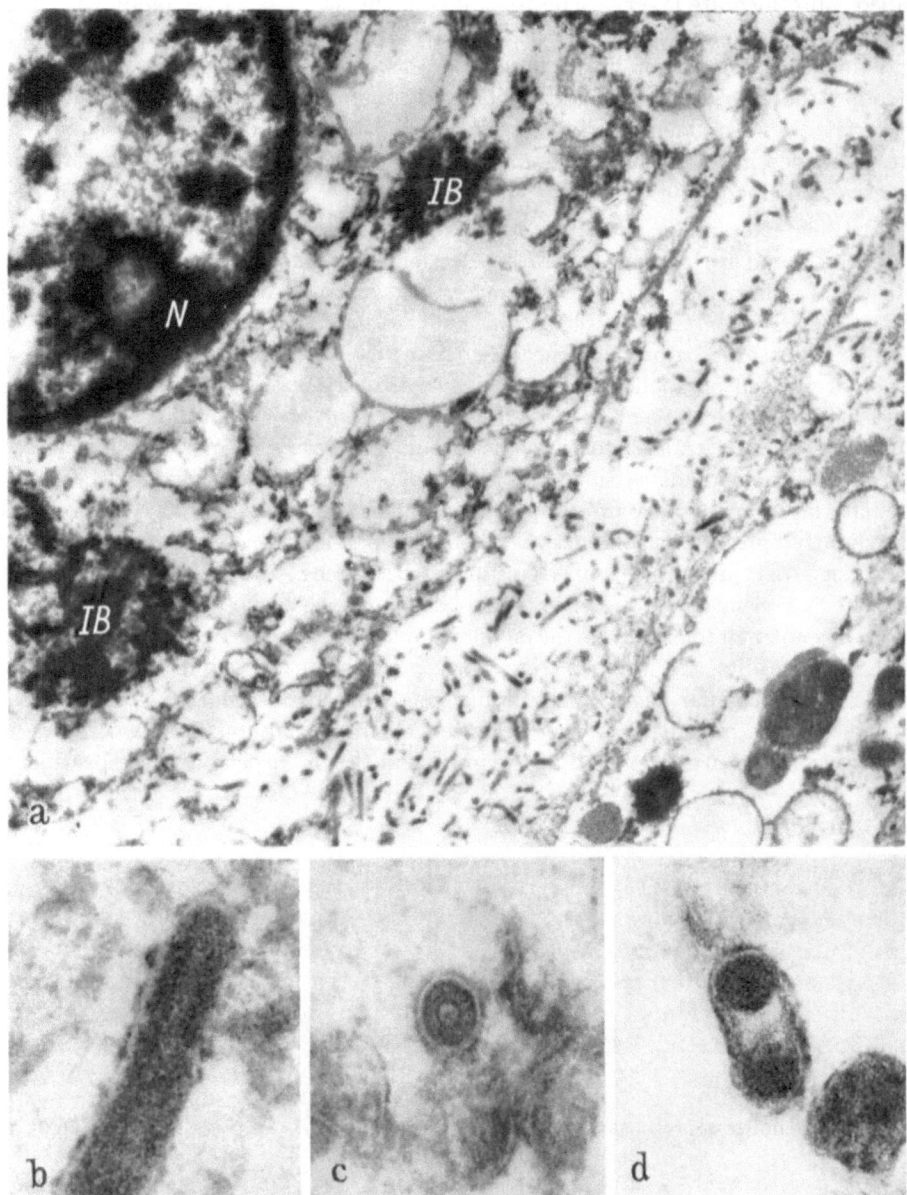

Fig. 3a—d. Marburg virus in *Cercopithecus aethiops* liver, 8 d p.i. a) Intercellular virions, mostly in the straight form; virus specific inclusion bodies (IB), nucleus (N), 15 000 ×; b) Longitudinal section; light axis, cross-striation, injured envelope; c) Cross section through a straight virion; axis with dot, surrounding material comprising nucleocapsid and intermediate layer, envelope; d) Cross section through coiled strand; envelope enclosing the coil. 150 000 ×

at sites where the outer envelope layer is thin or nonexistent (Fig. 2b). This observation indicates that at least one additional tubular layer with an outer diameter around 55 nm is located beneath the envelope. This substructure will be called "intermediate layer". The thickness of this layer is as yet unknown. The possibility that additional material is arranged between intermediate layer and nucleocapsid cannot be excluded. Occasionally in clusters of particles in blood, structures may be found which are completely devoid of the enclosing material, obviously identical with the nucleocapsid (Fig. 2c). Since these clusters are embedded in some matrix-like material, they may be related to developmental stages (see below).

In infected organs of one of the patients, of monkeys and guinea pigs, the agent has been found by means of ultrathin sections. The concentration in livers and spleens of guinea pigs in our experience is considerably smaller than in organs of *Cercopithecus aethiops*. In sections of infected monkey liver, huge amounts of virions are observed along with heavily damaged cells. In areas where the original order of the cell organelles still can be recognized, complete virions can only be discerned in extracellular positions (Fig. 3a). The overwhelming majority of the virions is in the straight form, some others exhibiting a more or less coiled form. In longitudinal sections the inner part of the virion including the intermediate layer appears as an electron-dense strand in which the 53 Å period and the slightly lighter axis zone may become visible (Fig. 3b). The envelope mostly shows signs of injury. Since embedding in water-soluble media (e.g. Durcupan) leads to better preservation of the virion, one can assume that the normal dehydration process is not well tolerated by the envelope material. Cross sections are in keeping with these results, the axis, of course, being much better visible (Fig. 3c). The rather thick tubular layer surrounding the axis should comprise the nucleocapsid and the intermediate zone. Indeed this material contains two layers of higher electron density, located at the inner and the outer periphery, respectively,

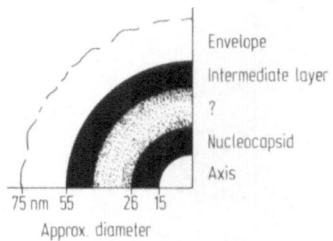

Envelope

Intermediate layer

?

Nucleocapsid

Axis

75 nm 55   26   15
Approx. diameter

Fig. 4. Schematic representation of the different layers of complete Marburg virus as seen in sections

which might represent these substructures. Yet more material of lower electron density lies in between. The axis area appears mostly empty, but small dots may occur (Fig. 3c). Occasionally entirely dense axial zones can be observed (Fig. 3d). Transverse sections through coiled virions in which the inner strand has been cut twice (circle type) or three times (spiral type) indicate that the envelope encloses the coil as a whole (Fig. 3 d). The observations on sections which are represented schematically in Fig. 4 are in good agreement with the results of negative staining.

Development

Studies on the sequence of developmental stages have been performed using infected cultures of Vero cells, a continuous kidney cell line derived from *Cercopithecus aethiops*, to which the virus had been adapted in three preceding passages. The infection was carried out at a multiplicity of about 10^{-1} ID_{50} per cell. Sections of the 8-hour material revealed neither virus particles of the inoculum nor any indication of the new virus generation. At 24 hours, the only alteration seen was accumulations of blurred wavy strands having a diameter of about 55 nm, partly fusing, appearing in the cytoplasm (Fig. 5). Since the number of cross sections is rather small, a lamellar arrangement of a part of this material must be considered. Although SELLERS staining and immunofluorescence at this time already demonstrate virus specific inclusion bodies, the above described accumulations need not necessarily be regarded as their ultrastructural equivalent.

Later on, at 32 hours after infection, more scattered, straight strands are seen, cross sections revealing the existence of the axis. From the cell surface complete virus particles exhibiting more density than the strands have extended into the extracellular space and into cytoplasmic vesicles. Straight forms as well as coiled and annular particles have appeared (Fig. 6 a). At some places rare examples of budding can be observed (Fig. 6 b). The cytoplasmic virus strands, according to their diameter, have to be regarded as the compound union of nucleocapsid and intermediate layer. They are not equipped with the envelope which seems not to be contributed until the release process. The pleomorphism typical of this virus may be caused by some unknown random events in the course of this phase; it indicates a rather high flexibility of the inner strand. Occasionally, however, areas of the cell-virus system may be observed where only long straight virions have emerged (Fig. 7). — No signs of a participation of the nucleus have been found.

The appearance of intracytoplasmic strands, first in loose accumulations and then scattered, and their completion during the release process must be regarded as the minimum events in the course of the virus development. Other manifestations of the infection are compact cytoplasmic inclusion bodies of a remarkable variability. They were observed in tissue culture from the 32nd hour after infection as well as in organs of infected animals. In the monkey liver material which originated from the late phase of infection, different forms of inclusion bodies were found, but severe cytopathic alterations had affected their preservation (Fig. 3). Better results were obtained when Vero cells and guinea pig liver were examined. Both materials presented a surprising variety of inclusion types, only some of which can be discussed here. Using consecutive thin and thick sections, it could be proved that the inclusion bodies seen both by light and electron microscopy are identical (Fig. 8). On the other hand it could be shown in the same way that certain elements in the cytoplasm which might be regarded as virus-specific inclusions were really micronuclei. In staining the thick Epon sections for light microscopy, good results were achieved by using the methods recommended

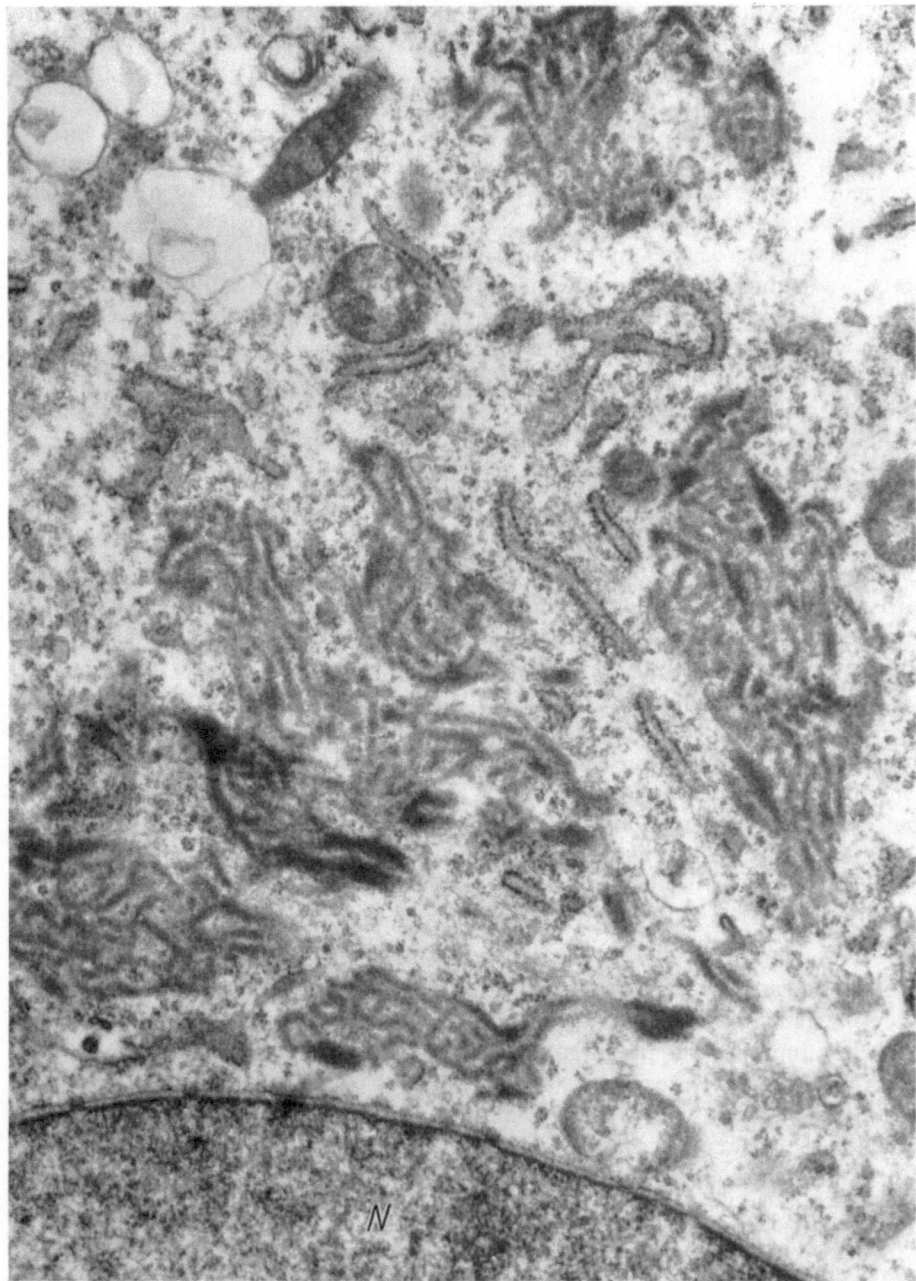

Fig. 5.   Vero cell, 24 h p.i.   Cytoplasmic accumulations of virus-specific material.
Nucleus (N).   24 000 ×

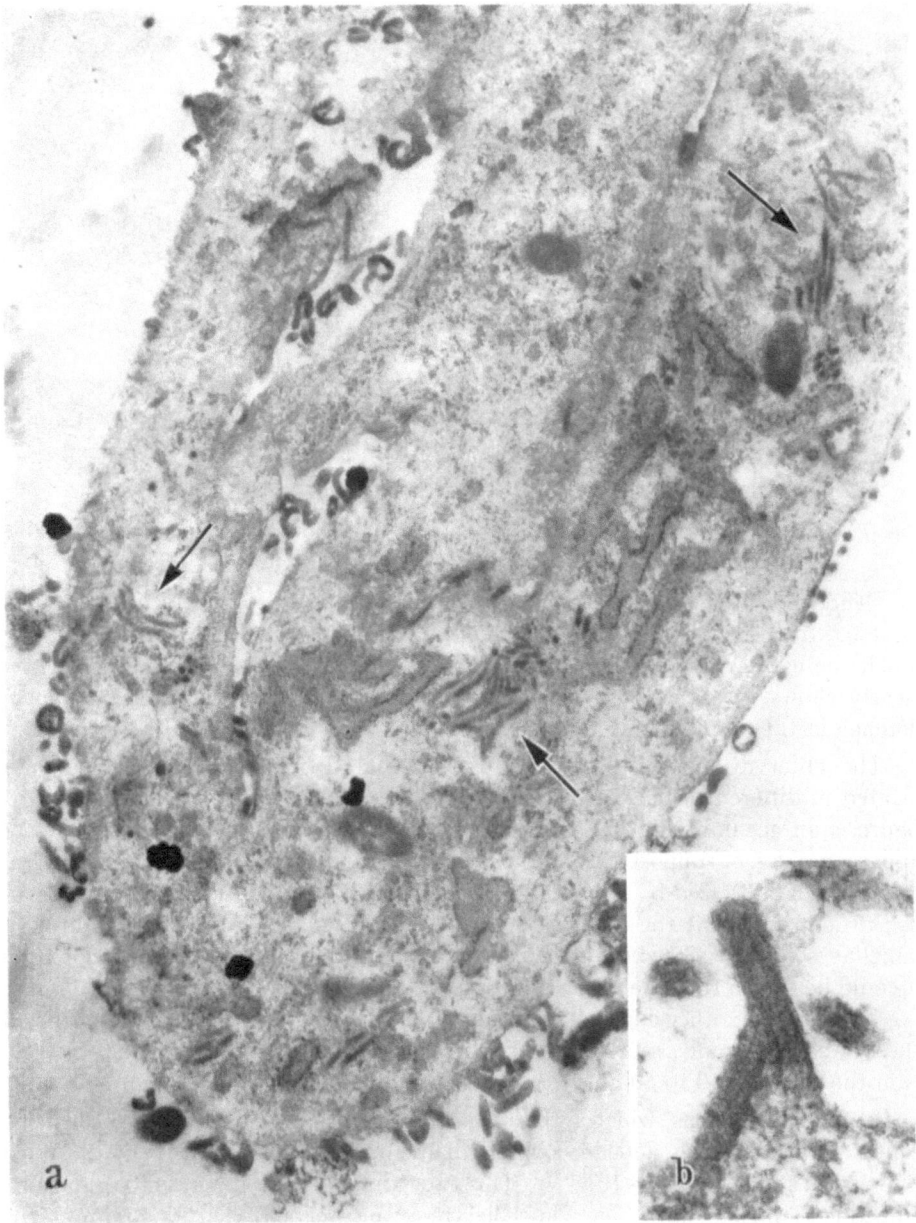

Fig. 6a and b. Vero cell, 32 h p.i. a) Scattered virus specific strands in the cytoplasm
(→) and complete virions on the cell surface and in cytoplasmic vesicles. 15 000×
b) Budding process at the cell surface. 100 000 ×

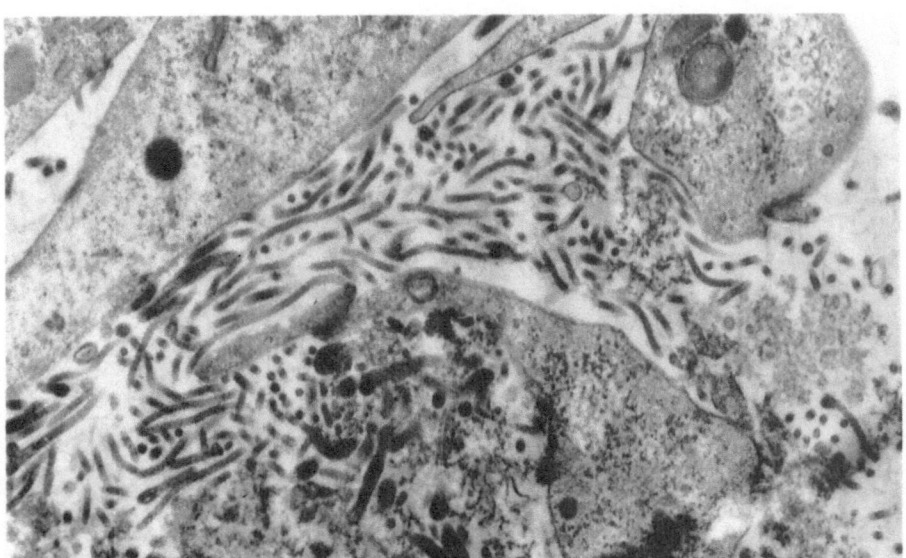

Fig. 7. Vero cells, 32 h p.i.   Straight virions located in an intercellular space.   15 000 ×

for the detection of Negri bodies (Paarmann, Muromzew, Sellers and Machia-vello).   In the light microscope the inclusion bodies of monkey livers stand out much more clearly than those in guinea pig livers.   The electron microscopy clearly shows that this phenomenon is due to the higher degree of cytoplasm damage in the monkey material.

The observation, represented by Fig. 8, that different inclusion types are located in different cells seems to be valid in general.   The smaller body in this figure is an accumulation of strands, while the larger one presents a spongelike appearance, a fact supported by serial sections.   In another type, not shown here, which is characterized by a concentric alignment of strands, serial sections estab-lished definitely that the observed "strands" are actually cross sections through lamellae, some of which show a palisade-like subdivision.   With the same technique it could be shown that the "granules" in inclusions like that shown in Fig. 14 may be spheres, or in other examples, cross sections through strands.   Common to all these lamellae, strands, and spheres is that their thickness or diameter, respectively, is in the range of 50 to 60 nm.

In other inclusions, which hitherto have only been observed in Vero cells, the center is stuffed with delicate-looking filaments and the periphery is occupied by heavily stained strands (Fig. 9).   This type and another one presenting matrix-embedded lamellae and spheres in a bizarre arrangement (Fig. 10), is reminiscent of similar manifestations in rabies-infected cells [14].   Further types like those shown in Fig. 11 have also been shown in cultures infected with the virus of vesicular stomatitis [15], and others, such as that represented in Fig. 12, resemble inclusions of the Egtved virus of the rainbow trout [16].

The two last mentioned formations clearly are regular patterns of tightly packed nucleocapsids which may or may not be embedded in a matrix of high

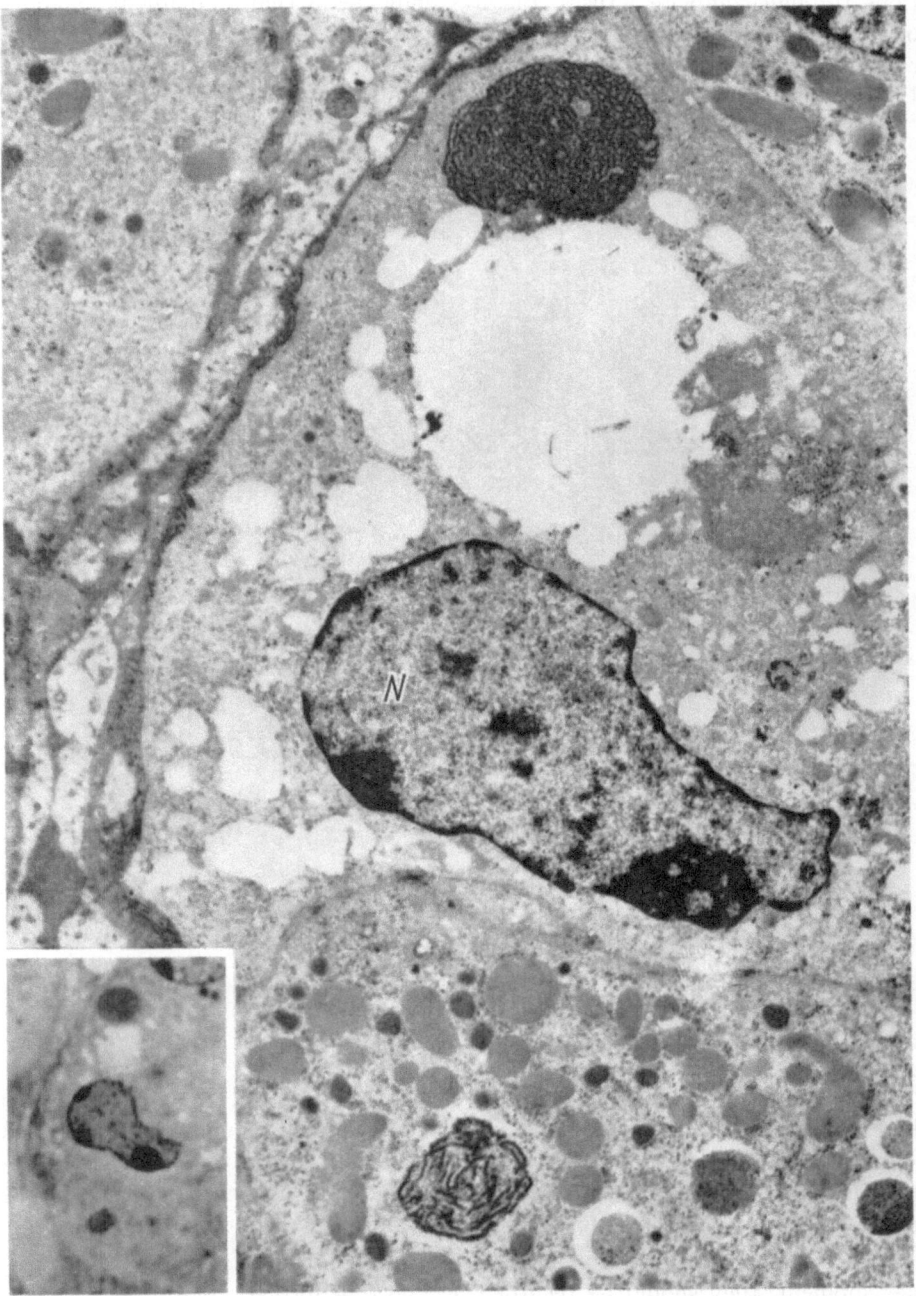

Fig. 8.   Infected guinea pig liver.   Two different compact inclusion bodies in two different cells.   Nucleus (N).   7000 ×.   Inset: Adjacent thick section for light microscopy of the same region; Muromzew-stained.   1400 ×

density. The central dot in the axis zone (Fig. 11), already known from cross sections through complete viruses (Fig. 3c), indicates a central filament of about 60 Å in diameter. In Fig. 13 parts of an inclusion are shown in which the nucleocapsids are embedded in a less tightly packed matrix. Cross sections, like that of Fig. 11, disclose a hexagonal pattern of the nucleocapsids, which have an outer

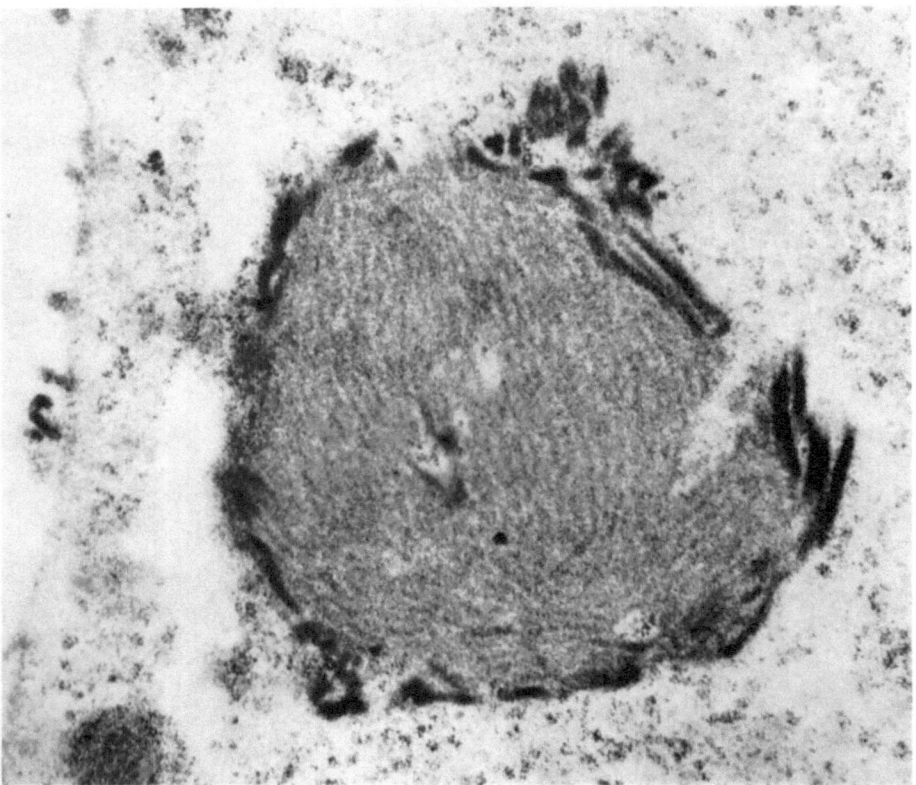

Fig. 9. Inclusion body in Vero cell, 3 d p.i., with delicate-appearing filaments in the center and dense strands in the periphery.   22 000 ×

diameter of about 26 nm, and longitudinal sections reveal the 53 Å cross-striation known from the results of negative staining. At higher magnification a beaded substructure of the nucleocapsid may be seen, pointing to the existence of approximately 24 grains, less than 50 Å in diameter, in the circumference of the nucleocapsid (Fig. 13a, inset).

Since the compact inclusion bodies do not represent integral stages of the virus development, they should be regarded as abortive formations of surplus virus specific material. It can be assumed that they are derived from the loose accumulations of strands observed during the early stages of the virus development. Their variations in appearance might be the outcome of competing processes of development, morphogenesis, and lytic degeneration.

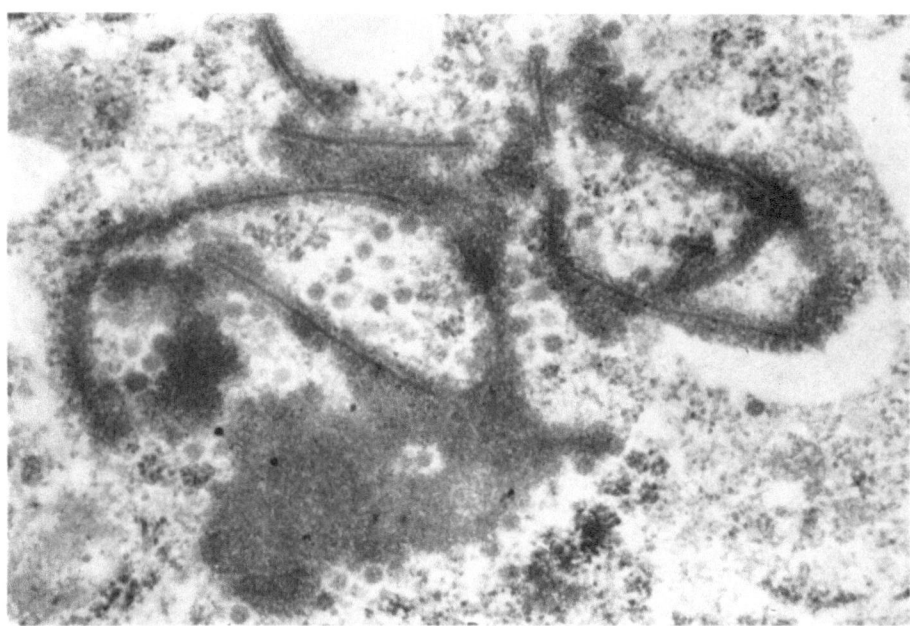

Fig. 10. Bizarre inclusion body in guinea pig liver, showing lamellae embedded in matrix, and spheres.   40 000 ×

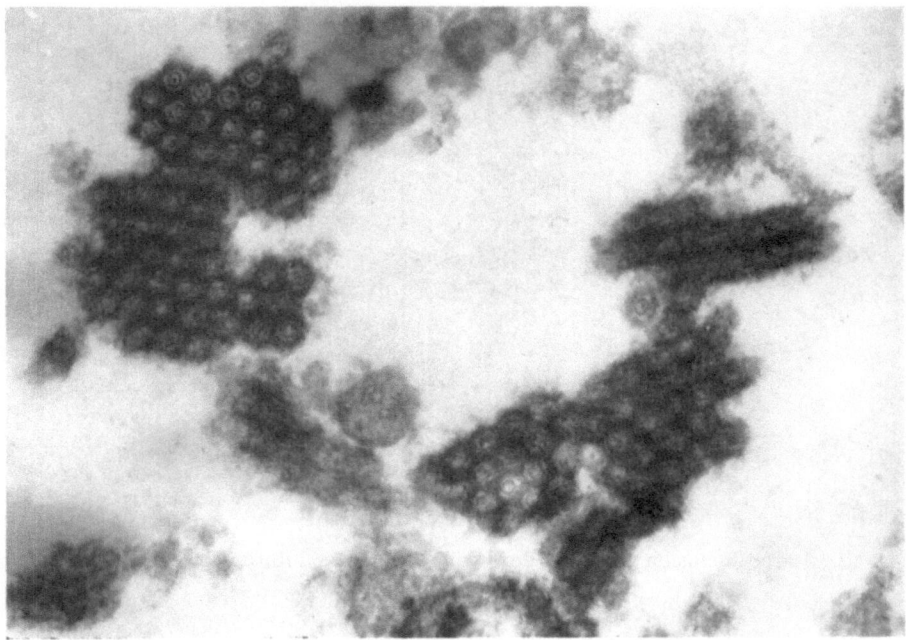

Fig. 11. Dense inclusion body in Vero cell, 5 d p.i. showing a regular pattern of nucleo-capsids embedded in matrix, partly equipped with an axial filament.   80 000 ×

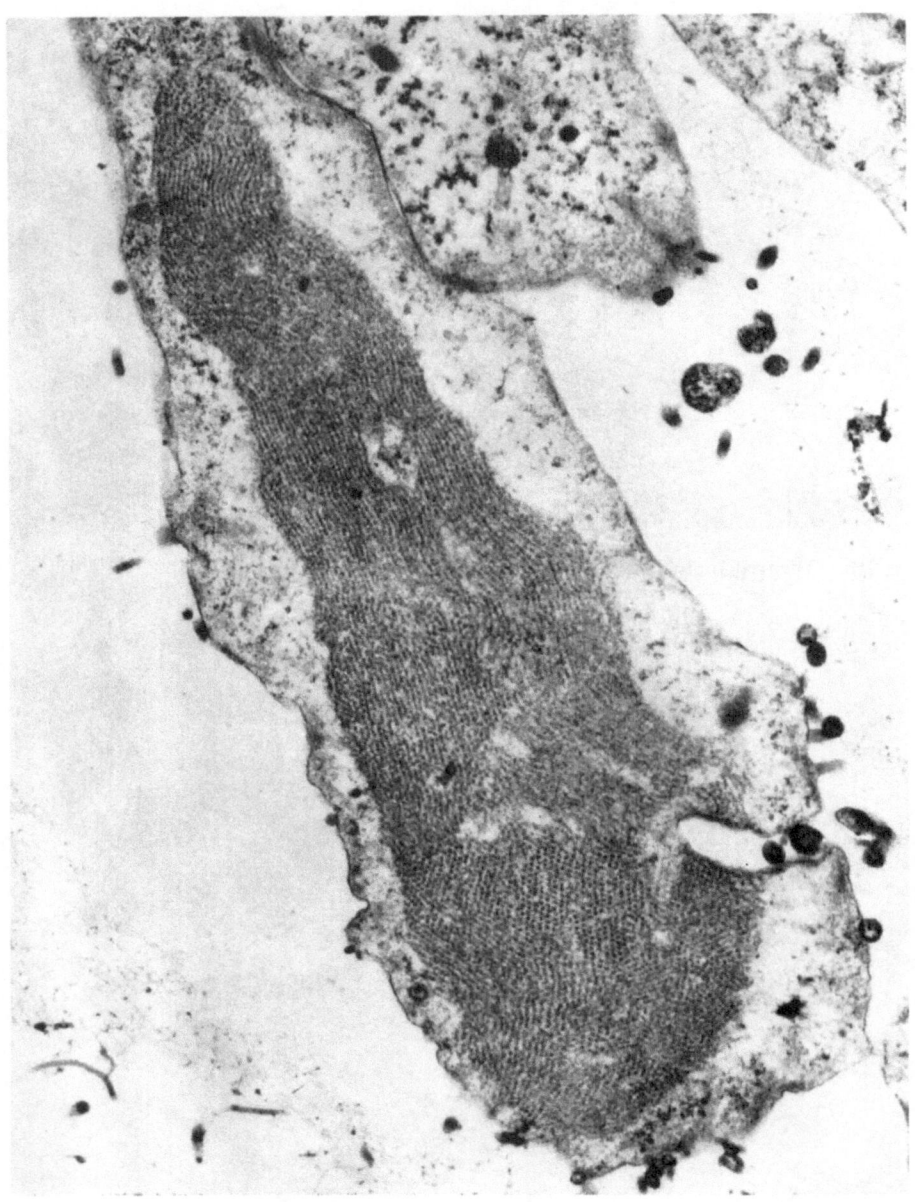

Fig. 12. Large inclusion body in Vero cell, 3 d p.i., exhibiting ordered assembly of naked nucleocapsids.   20 000 ×

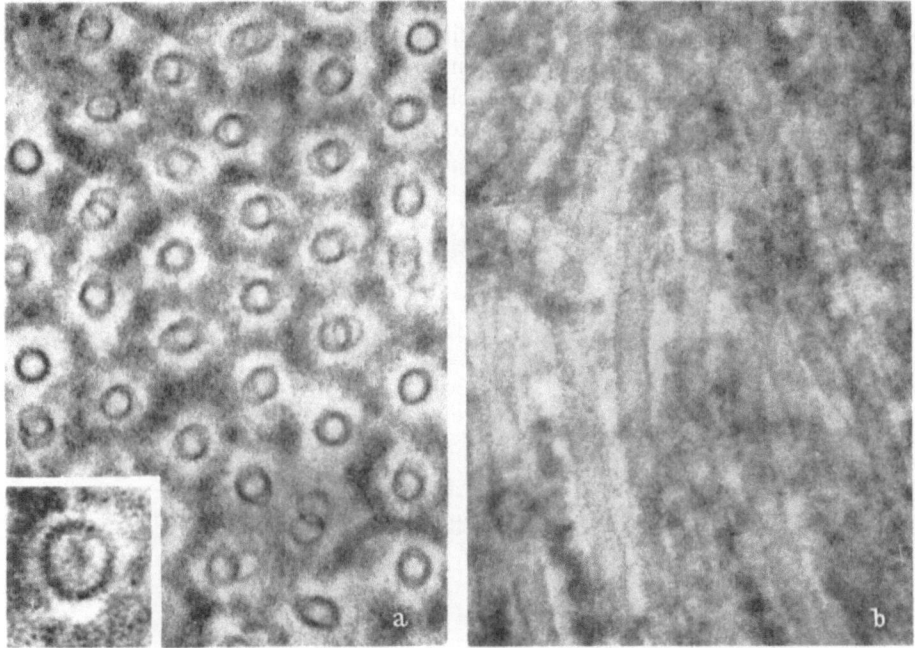

Fig. 13a and b. Portions of inclusion bodies in Vero cells, 5 d p.i., showing nucleo-capsids in loose arrangement of matrix. a) Cross section. (Inset: Beaded structures in the circumference of a nucleocapsid.   400 000 ×); b) Longitudinal section, showing the 53 Å cross-striation.   180 000 ×

Classification

It has been reported that bromodeoxyuridine [11] and actinomycin [17], which interfere with the replication of DNA, do not inhibit the synthesis of the Marburg virus antigen. This finding is supported by ultracytochemical results. The applica-tion of the DNA-specific HCl/silver-methenamine reaction [18] demonstrates heavy silver deposits at the sites of chromatin, but no reaction with the inclusion bodies (Fig. 14) and virus particles. On this basis it can firmly be concluded that the genetic material of this virus is RNA.

The morphological results presented here point out that a close relationship exists between the Marburg virus and a group of RNA viruses typified by the viruses of vesicular stomatitis and rabies. Different designations have been proposed for this group: "Stomatoviridae" [19], "Rabies viruses" [20], "Bullet-shaped viruses" [21]. Yet, apparently the term "Rhabdoviruses" [22] is used most of all, though it must be borne in mind, that this name has also been applied to naked helix RNA viruses of plants [19, 23]. Other members of this group are the Flanders-Hart Park virus, the Kern Canyon virus, the Sigma virus of Droso-phila, the Egtved virus of trout, and several viruses causing diseases in plants.

The supposition that some relationship might exist between the Marburg virus and a highly ordered strand-like structure associated with leptospires [25] had to be abandoned because of unequivocal differences in morphology [26].

Almost all structural details of the Marburg virus, as for example the existence of two cross-striations of different periodicity, the spike-like projections, the beaded substructure of the nucleocapsid and even the branching, have their counterparts somewhere in the work of others describing viruses of this group. The same is true of the developmental stages and, to a certain degree, also with regard to the inclusion bodies.

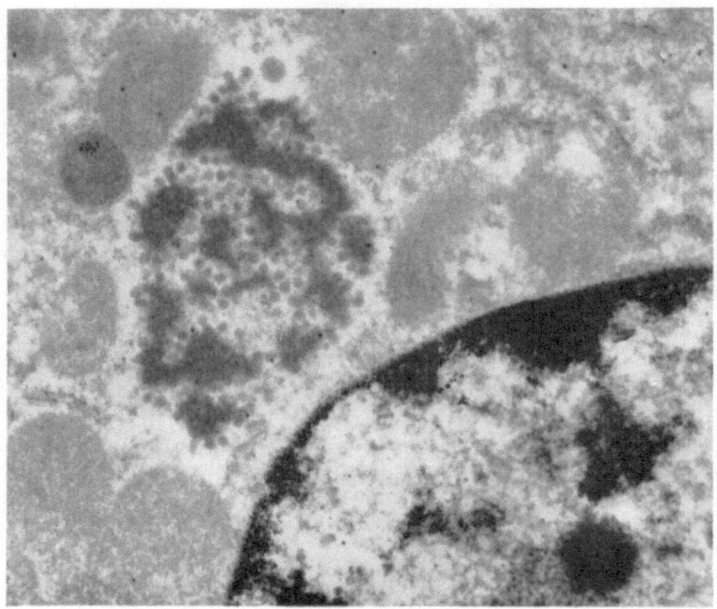

Fig. 14. Inclusion body with matrix and spheres in guinea pig liver. Treatment of section with silver methenamine solution after HCl-hydrolysis shows DNA content of chromatin but no reaction of the inclusion body. Counterstained with uranylacetate.
25 000 ×

Although unusually long particles have been described occasionally in work on other viruses of this group, there is no doubt that the great length of the Marburg virus particles, which in the normal virion is about three times that of the other viruses, is exceptional. The coiled or annular forms typical of this virus can be understood as a consequence of flexibility and great length, but this feature, too, has been observed in other rhabdoviruses [27, 28]. Bullet-shaped forms have never been seen in our material, but two rounded ends, as in the Marburg virus, have repeatedly been described in others.

As the tendency to develop to great length cannot be regarded as a fundamental difference, the Marburg virus should be included into the rhabdovirus group.

We thank Prof. R. Haas and Dr. G. Maass, Freiburg, for the monkey material and Miss B. Mill, Miss A. Stromeyer, Mrs. Wolff and Mr. H. Giese for technical assistance.

References

1. HENNESSEN, W., BONIN, O., MAULER, R.: Dtsch. med. Wschr. **93,** 582 (1968).
2. HAAS, R., MAASS, G., OEHLERT, W.: Med. Klinik **63,** 1359 (1968).
3. SMITH, C. E. G., SIMPSON, D. I. H., BOWEN, E. T. W., ZLOTNIK, I.: Lancet **1967 II,** 1119.
4. SIEGERT, R., SHU, H.-L., SLENCZKA, W.: Dtsch. med. Wschr. **93,** 604 (1968).
5. MAY, G., KNOTHE, H.: Dtsch. med. Wschr. **93,** 620 (1968).
6. SIEGERT, R., SHU, H.-L., SLENCZKA, W., PETERS, D., MÜLLER, G.: Dtsch. med. Wschr. **92,** 2341 (1967).
7. PETERS, D., MÜLLER, G.: Dtsch. Ärzteblatt **65,** 1831 (1968).
8. HAAS, R., MAASS, G., MÜLLER, J., OEHLERT, W.: Ztschr. Med. Mikrobiol. Immunol. **154,** 210 (1968).
9. MAY, G., KNOTHE, H., HÜLSER, D., HERZBERG, K.: Zbl. Bakt. I. Orig. **207,** 145 (1968).
10. KUNZ, CH., HOFMANN, H., KOVAC, W., STOCKINGER, L.: Wien. Klin. Wschr. **80,** 161 (1968).
11. KISSLING, R. E., ROBINSON, R. Q., MURPHY, F. A., WHITFIELD, S. G.: Science **160,** 888 and **161,** 1364 (1968).
12. ZLOTNIK, I., SIMPSON, D. I. H., HOWARD, D. M. R.: Lancet **1968 II,** 26 and 458.
13. MÜLLER, G.: Arch. ges. Virusforschg. **27,** 339 (1969).
14. HUMMELER, K., KOPROWSKI, H., WIKTOR, T. J.: J. Virol. **1,** 152 (1967).
15. DAVID-WEST, T. S., LABZOFFSKY, N. A.: Arch. ges. Virusforschg. **24,** 30 (1968).
16. ZWILLENBERG, L. O., JENSEN, M. H., ZWILLENBERG, H. H. L.: Arch. ges. Virusforschg. **17,** 1 (1965).
17. SLENCZKA, W.: Zbl. Bakt. I. Ref. **215,** 545 (1969).
18. PETERS, D.: Electron Microscopy 1966, Tokyo, Vol. II, 195.
19. Provisional Committee for Nomenclature of Viruses, "Proposals and recommendations". Ann. Inst. Pasteur **109,** 625 (1965).
20. BELL, T. M.: Arch. ges. Virusforschg. **18,** 257 (1966).
21. ANDREWES, C., GIBBS, A. J.: Int. Virol. **I** (Helsinki), 195 (1969).
22. MELNICK, J. L., McCOOMBS, R. M.: Progr. Med. Virol. **8,** 400 (1966).
23. HIRTH, L.: C. R. Acad. Sci. **261,** 4556 (1965).
24. KITAOKA, M., MURPHY, F.: Int. Virol. **I** (Helsinki), 309 (1969).
25. ALMEIDA, J. D., WATERSON, A. P., BERRY, D. M., TURNER, L. H.: Lancet **1969 I,** 235.
26. PETERS, D., MÜLLER, G.: Lancet **1969 I,** 923.
27. ATANASIA, P.: Path.-Biol. **15,** 1103 (1967).
28. ARSTILA, P., HALONEN, P., SALMI, A.: Arch. ges. Virusforschg. **27,** 198 (1969).

Morphology and Morphogenesis of the Marburg Agent

J. D. ALMEIDA, A. P. WATERSON, and D. I. H. SIMPSON

With 21 Figures

Introduction

It is now generally recognised that the causal agent of the Marburg monkey disease is a virus which is morphologically unlike any known virus (SHU, SIEGERT and SLENCZKA, 1968; MARTINI, 1969). The appearance of the agent has been described in thin sections by PETERS and MÜLLER (1968), KISSLING et al. (1968) and ZLOTNIK et al. (1968) and in negatively stained preparations by SIEGERT et al. (1967), PETERS and MÜLLER (1968; 1969a), KISSLING et al. (1968), KUNZ et al. (1968), and MAY et al. (1968). ALMEIDA et al. (1969) published a description of structures associated with leptospires which bore a considerable resemblance to those described for the Marburg agent. However, the precise relationship between these structures and the Marburg agent has remained obscure (PETERS and MÜLLER, 1969b) and it seemed that further information on the morphology of the Marburg agent was needed before any conclusion could be drawn.

We have undertaken such a study mainly on material propagated in BHK21 cells, and have found that three principal morphological forms of the Marburg agent can be identified. These are a naked helix; a sinuous, probably immature form; and a third, which we consider the mature form of the agent. The last is most frequently seen in the form of a doughnut which in mathematical terminology is referred to as a torus.

In addition to purely morphological studies it also seemed essential to establish that the particles visualised in the electron microscope are in fact the agent of the disease. Using immune electron microscopy it has been possible to show that specific antisera, either from convalescent human patients, or produced in animals, coat the particles of the agent associated with the Marburg monkey disease with antibody molecules, while control sera leave them unchanged.

Materials and Methods

Morphological studies on the Marburg agent were carried out as follows. Virus which had been passed nine times in guinea-pigs and three times in rhesus monkeys was passed six times in BHK21 cells. Six days after infection the cells were harvested and spun down at low speed, the supernatant being discarded. These cells were then lysed with distilled water and left for one hour so that maximum disruption would take place. Formalin to a concentration of 0.2% was then added so that the material could be handled with safety for electron microscopy. The cell suspension was concentrated by centrifuging for 30 minutes at 10,000 r.p.m. in a Sorvall MS2 centrifuge, rotor 425, and the pellet suspended in approximately one tenth of the original volume of distilled water. Negative staining was carried out by mixing a drop of this material with an equal quantity of 3% phosphotungstic acid adjusted to pH6. Using a Pasteur pipette a drop of this mixture was placed on a 400 mesh carbon formvar coated grid and excess fluid withdrawn. When dry the grid was examined immediately in a Philips 300 electron microscope.

The same six day material from BHK21 cells was also used for the immune electron microscopy. However, in order to make the virus suspension suitable for the addition of antiserum it was found necessary to eliminate all traces of formalin. This was done by three cycles of centrifugation of 30 minutes at 10,000 r.p.m. as above. Each time, the pellet was resuspended in the original volume of phosphate buffered saline and the third such suspension was used for mixing with serum. Antiserum had been obtained from two human cases and also from infected guinea-pigs and hamsters. The experiment was controlled with normal human and rabbit sera. In every case 1.0 ml of the virus suspension was thoroughly mixed with 0.1 ml of the sera to be tested. The mixture was left for one hour at 37 °C and then overnight at 4 °C. The following morning the preparations were spun for 30 minutes at 10,000 r.p.m. and the pellet used as previously described for negative staining.

Results

Negatively stained preparations of the Marburg agent revealed three distinct forms:

1. A naked helix
2. A sinuous membrane-covered form
3. The mature form of the virus, predominantly doughnut-like, (i.e. a solid ring).

Intermediate forms between all three of these groups have been observed, allowing us to describe them and also to suggest the association between them.

The three categories will be described separately:

1. The Naked Helix

This structure was found either singly, or in larger groups when associated with cell material (Figs. 1, 2). It is a linear structure of varying length but with certain constant features. The most outstanding of these in a sharply defined central core with a diameter of 280 Å. The helix itself gave an impression of standard size but in almost every case the outer edge appeared frayed, making measurements difficult and giving a wide size range of 500—800 Å. The pitch also appeared variable and was again dependent on the state of preservation of the helix, an average value being 85 Å. Intermediate forms revealed that this uncovered helix acquires a membrane and becomes the internal component of the sinuous form next to be described (Fig. 3).

2. Sinuous Particles

This form of the virus is by far the most pleomorphic of the three and indeed has to be broken down into three sub-groups, which although sharing certain basic features, differ sufficiently from each other to be described separately. They will be designated *covered*, *full*, and *empty*.

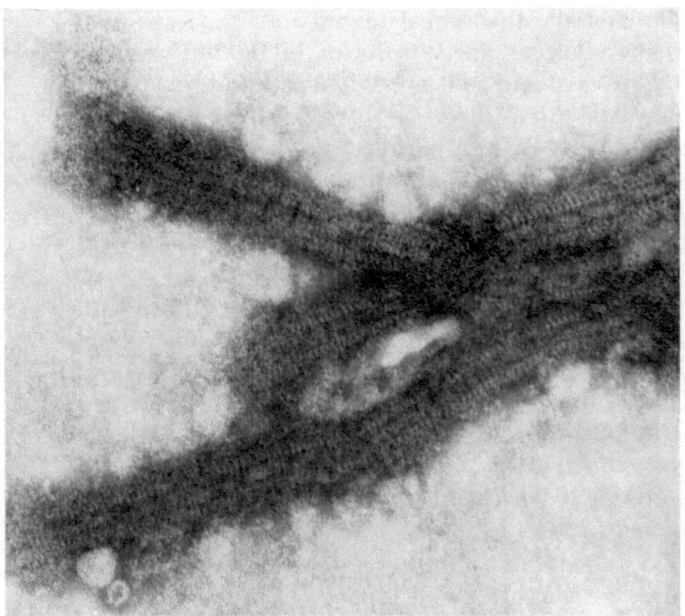
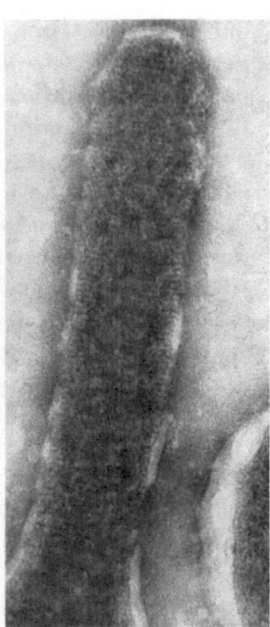

Fig. 1   Fig.2

Fig. 1. A group of naked helices showing the distinctive 280 Å core. The outer surface of these structures appears ragged, making it difficult to obtain accurate measurements. This component of the Marburg agent bears a resemblance to the internal helices of some RNA viruses.   × 105,000

Fig. 2. A single membrane-bound particle showing an unusual view of the internal helix. In this case it is the exterior of this component rather the core which has been delineated by the stain. The diameter of the helix is approximately 390 Å.   × 170,000

a) Covered sinuous particles

These were found frequently in the preparations and appeared as long linear structures occasionally several μ in length (Fig. 4). Except for distinctive projections, approximately 100 Å in length, no detailed structure was visible. There is considerable variation in diameter for particles of this group, the range being 720—1100 Å. These measurements exclude the projections, and all diameters will be given in this way. Intermediate forms revealed that, when degraded, the covered sinuous particle contained the structures that are described in the next two categories, i.e. the full and empty sinuous forms (Fig. 5).

b) Full sinuous particles

As with the covered sinuous form, projections were present on the surface, but it was also possible to see that these were attached to a membrane of variable thickness, having no regular structure. This membrane, both in position and appearance, resembles the projection-bearing outer envelope of many compound viruses. It should also be remarked that several of these sinuous particles had an arrangement of this outer, probably lipoprotein, membrane resembling that found in the structures associated with leptospires and indeed that of the leptospires themselves (Figs. 6—8). Inside this membrane a sheath with a periodicity of 45 Å

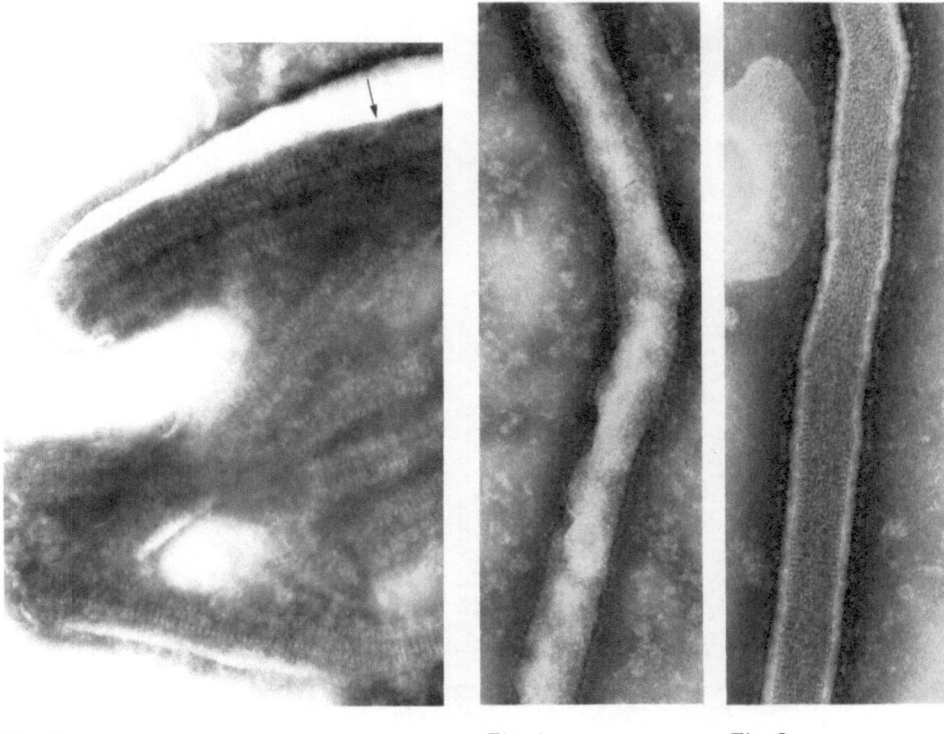

Fig. 3   Fig. 4   Fig. 5

Fig. 3. Another group of internal helices showing intermediates between naked and membrane-bound. It is of interest that at the top right of the micrographs (arrow) the 45 Å periodicity of the outer sheath can be just seen.   × 100,000

Fig. 4. A covered sinuous particle illustrating the distinctive projections, approximately 100 Å long, on the surface. Since the average diameter of this tubule is 700 Å it represents an empty sinuous form similar to the example in Fig. 5.   × 100,000

Fig. 5. A tubule similar to that shown in Fig. 4, but sufficiently degraded to allow visualisation of the internal structure at the lower end. Here it can be seen that below the outer projection-bearing membrane lies a sheath with a 45 Å periodicity. × 125,000

could be seen, while, within this structure again, there was an internal helix of 390 Å with a central core of 280 Å and a pitch of 78 A (Figs. 9, 10). Most frequently the outstanding feature of this inner helix was the 280 Å inner core, but occasionally staining conditions were such that it was possible to visualise the outer surface of this component (Fig. 2). The overall diameter of this form is close to 1000 Å, that of the periodic sheath 780 Å. This construction is shown diagrammatically in Fig. 21a.

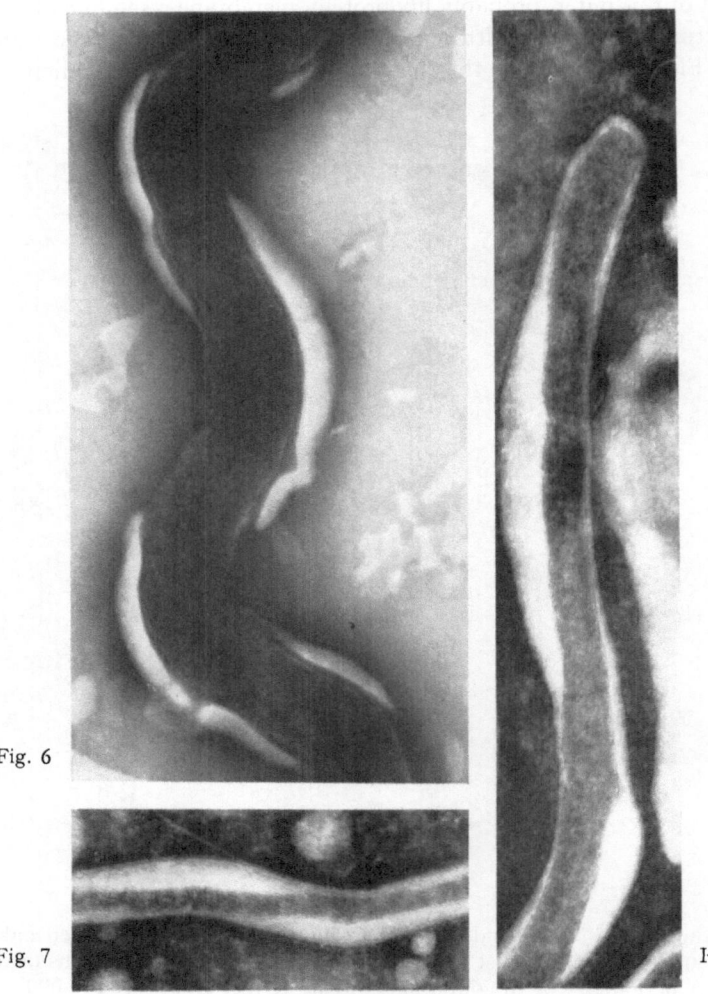

Fig. 6

Fig. 7

Fig. 8

Fig. 6. A low power micrograph of a leptospire showing that this organism frequently displays thickening of its outer membrane on the convexities of its spiral.   × 95,000

Fig. 7. A leptospire-associated structure showing an arrangement of outer membrane similar to that of the leptospire itself.   × 115,000

Fig. 8. An example of an empty sinuous particle of the Marburg agent, again showing an arrangement of the outer membrane suggestive of a spiral.   × 115,000

c) Empty sinuous particles

Like the full sinuous this form also displayed projections, although it should be remarked that when these forms were considerably degraded the projections were no longer visible. It also displayed a layer of amorphous membrane beneath the projections and inferior to this the periodic sheath with 45 Å spacing (Fig. 11). However, no structure internal to this could be resolved and it was also found that this form was considerably narrower than the full sinuous, having an overall diameter of approximately 720 Å and an internal measurement of 500 Å for the periodic sheath (Fig. 21 b). Finally, examples were found, (Figs. 12, 13) showing that this form is rightly described as empty, for, as can be seen, the micrograph shows a sinuous form displaying, for half its length, an internal helix, and, after a constriction, a narrower empty portion having no internal helix.

All of these sinuous forms showed extreme flexibility, frequently forming loops. In addition branching was present and many tubules terminated in a bulbous swelling.

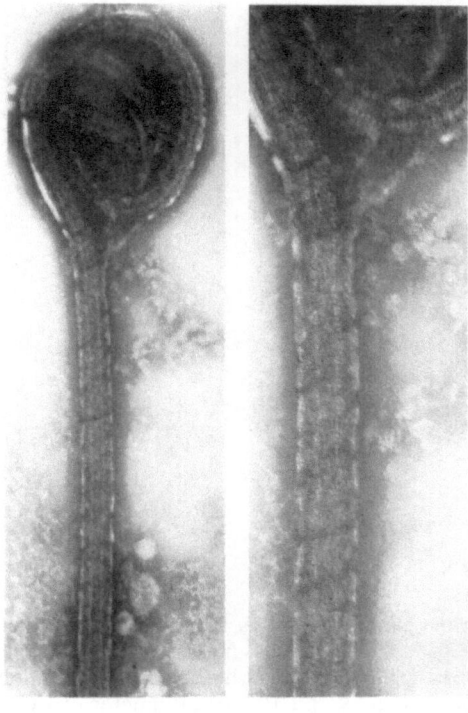

Fig. 9   Fig. 10

Fig. 9. An example of what we have termed the full sinuous form of the Marburg agent. Here the tubule formed from an outer sheath of 45 Å periodicity and an inner helix of 78 Å pitch, coil into a circular form at the end. Greater detail of this particle can be seen in Fig. 10.   × 56,000

Fig. 10. A high power of the central portion of the particle shown in Fig. 9. At this magnification it is possible to see the two different periodicities present in the structures forming the tubule.   × 100,000

3. Mature Particles

In all specimens examined there were particles, frequently in large aggregates, which contained all the elements of the full sinuous forms, but whose configuration was more compact (Fig. 14). The most common morphology was that of a torus or ring (Fig. 17); variations, such as open rings and bretzel forms, were also present (Fig. 15).　As with the sinuous forms, many of the particles were not penetrated by the stain, but a suggestion of an inner core, approximately 200 Å

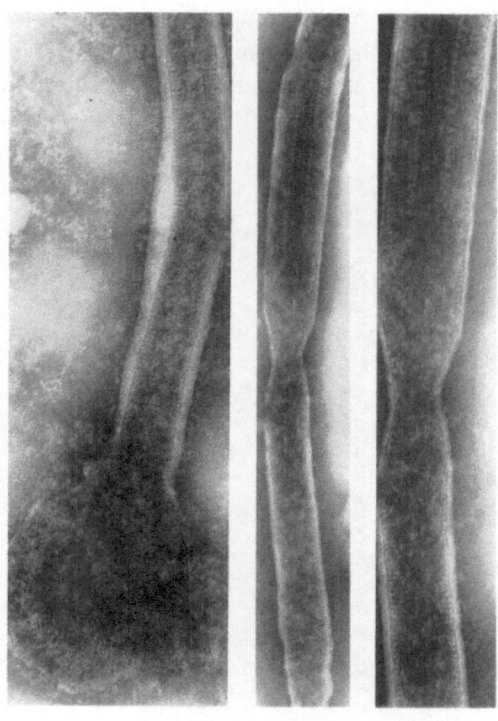

Fig. 11　　　　　　Fig. 12　　Fig. 13

Fig. 11.　A typical empty sinuous form is illustrated here.　Unlike the full sinuous particle illustrated in Figs. 9 and 10 there is no visible helix within the sheath with 45 Å periodicity.　In addition, this periodic sheath has a diameter of approximately 500 Å whereas the one in Figs. 9 and 10 has a diameter of over 700 Å.　× 125,000

Fig. 12.　An intermediate particle showing the relationship between the full and empty sinuous particles.　It can be seen that the upper portion of the micrograph contains an internal helix, while, after a constriction, the lower part does not.　When the internal helix is present the 45 Å periodic sheath has a diameter of approximately 700 Å, while the empty lower portion has a diameter of only 500 Å.　× 100,000

Fig. 13　A higher power of a portion of Fig. 12 showing the transition between the 'full' and 'empty' portion of particle.　× 300,00

in diameter, was almost invariably present (Fig. 15).　In others, the internal helix could be seen, just as in the "full" sinuous forms (Fig. 16).　Again, the helix displayed a core of 280 Å and it seems reasonable to suppose that this corresponds to the 200 Å core of the non-penetrated particles, the difference being due to the different penetrating power of the stain in the two forms.　The one outstanding feature of this (mature) form of particle was that the ring was interrupted by a varying number of constrictions, and in particles with visible internal helix it could be seen that the constriction corresponded with a break in this helix, so that each ring form contained a variable number of fragments of helix within it, the

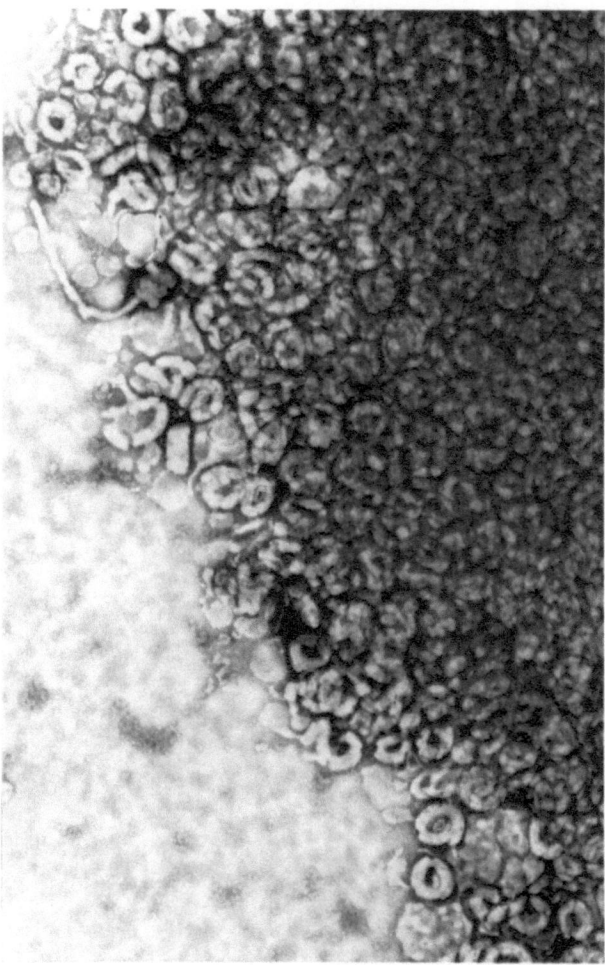

Fig. 14. A low power micrograph of a large aggregate of the particles that we consider the mature form of the Marburg agent. At this magnification the most striking feature is the doughnut or torus form of the majority of particles.   × 31,000

most common number being four. The outside diameter of these forms was 3000—4000 Å. Not infrequently the centre of these forms was imperforate, so that the membrane covered the whole structure, giving the form of a mammalian red cell, i.e. approximately a bi-concave disc. It should be remarked that only the full sinuous particles seem to give rise to the mature form, as the diameter of the tubule bent to form the ring was always close to 1000 Å, and nearly all of them reveal the core of the inner helix.

IMMUNE ELECTRON MICROSCOPY

The results of these experiments were clear cut and uncomplicated. Virus particles reacted with control sera showed no change in their morphology while those treated with test sera (human, guinea-pig, and hamster) were obscured by

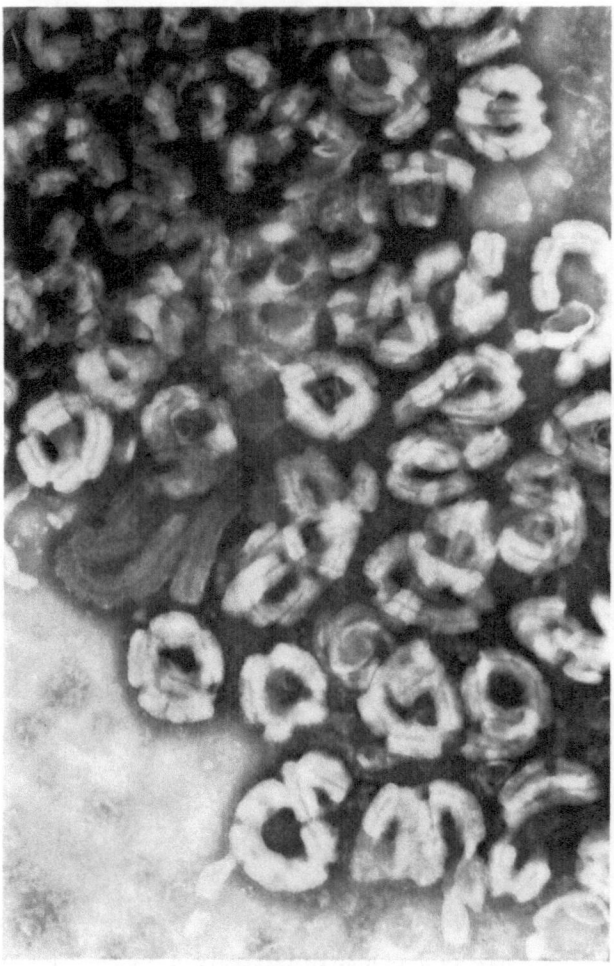

Fig. 15. A higher power of another aggregate of mature particles. Most of the particles have not been penetrated by the stain and the only feature visible is a 200 Å core which almost certainly corresponds to the 280 Å core of the internal helix. One particle which is shown in greater detail in Fig. 17 has been penetrated by stain and reveals the internal structure.   × 75,000

a distinct layer of antibody molecules (Figs. 18—20). The appearance of this phenomenon is now well established and has been described for several other virus systems (ALMEIDA and WATERSON, 1969).

Discussion

From the studies of previous workers two main structural types have emerged for the Marburg agent. These are what we have described as the sinuous forms and the circular doughnut forms (Figs. 9, 10, 15). The specimens that we have been able to examine contained the agent to a high titre, at least as concerns

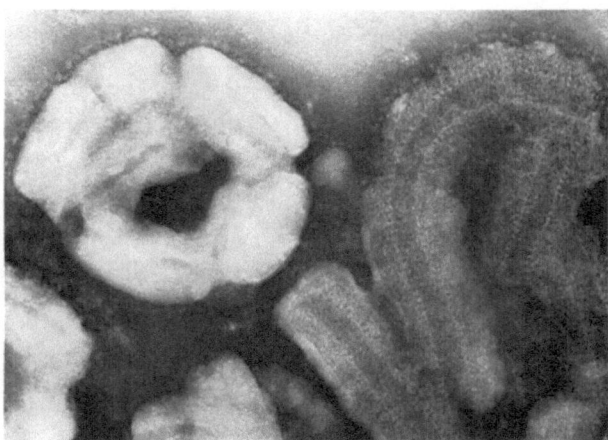

Fig. 16. A higher power of a small area of Fig. 15 showing the contrast between a particle that had, and one that had not, been penetrated by the stain. The particle on the right shows the disticntive constrictions that on a penetrated particle reveal that the internal helix is present in the form of short irregular segments.   × 200,000

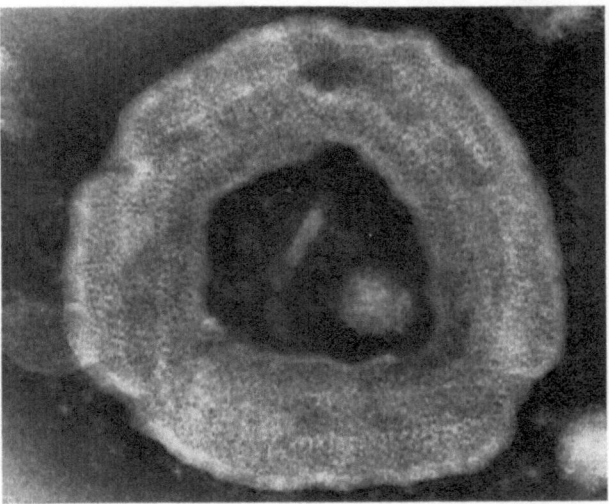

Fig. 17. A mature particle illustrating the torus form that leads us to suggest that the Marburg agent be made the founder member of a new group called the toroviruses. It is just possible to visualise the internal component within this particle.   × 300,000

physical particles, and this has allowed us to visualise not only these principal forms of the agent but also several intermediate ones. For example, it is now apparent that while we were able to visualise three morphologically different types of sinuous particle, the covered type was merely an uninterpretable example of one or other of the two genuinely different types of sinuous form present, one containing an internal helix, and presumably having the potential of viability, and the other, lacking this internal component, and therefore to be looked on as

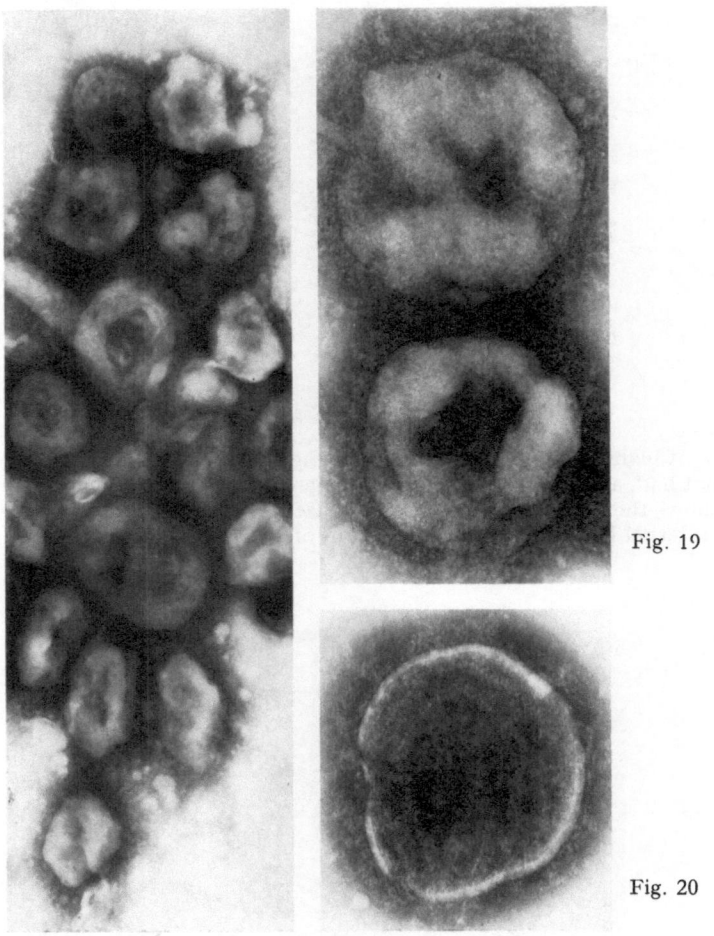

Fig. 19

Fig. 18   Fig. 20

Fig. 18. A group of the particles associated with the Marburg monkey disease aggregated by antibody present in a human convalescent serum. The particles are clumped together and, as can be seen at higher magnification (Fig. 19), obscured by the presence of antibody molecules. × 80,000. The fact that only specific antisera produced this effect makes it likely that these particles are indeed the agent associated with the disease.

Fig. 19. Two particles thickly coated with antibody molecules. The serum in this case was obtained from infected guinea-pigs.   × 150,000

Fig. 20. A single virus particle coated with antibody obtained from an infected hamster. Although obscured by the presence of antibody it is still possible to see the internal structure of this particle.   × 150,000

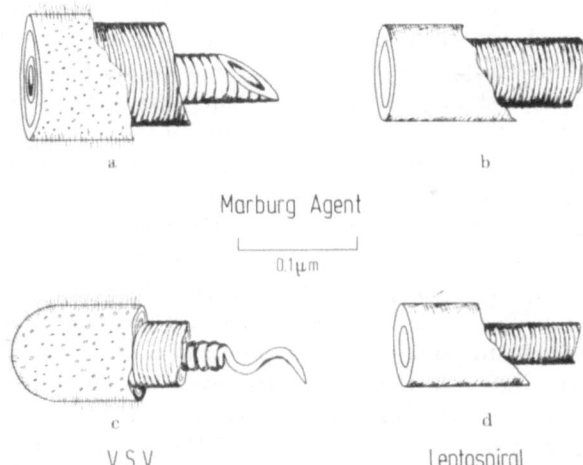

Marburg Agent

0.1 μm

V. S. V.   Leptospiral

Fig. 21 a — d. Diagramatic representations of relevant structures. a) Full, i.e. complete, particle of Marburg agent; b) Incomplete or empty particle of Marburg agent; c) Vesicular stomatitis virus, representative of the rhabdoviruses; d) Periodic structure found in association with leptospires.

an empty or incomplete particle. Again, both forming and degraded particles show that the naked helix (Fig. 3) is itself the internal component both of the full sinuous form (Figs. 9, 10) and the mature circular particle or torus (Fig. 16). It should be remarked here that the one constant measurement linking the naked helix, the full sinuous form and the mature particle, is the 280 Å diameter of the core of the internal helix. From these facts, even though the electron microscope is by no means a dynamic instrument, a possible means of assembly begins to present itself. Although we cannot be certain, the internal helix is most likely to be the ribonucleoprotein (RNP) of the virus. The groups of naked helices found in association with cell material bear a considerable resemblance to areas of RNP helix of viruses such as measles (WATERSON, 1965), even though larger in size. In addition this component occupies the protected internal location usually accorded by viruses to their nucleoprotein (WATERSON, 1968). We suggest then that the naked helix is the primary physical component of the virus and that this is surrounded by the sheath with the 45 Å periodicity. Finally, this whole structure is completed by the addition of a lipid-containing membrane bearing projections on the surface. The final stage of virus maturity is reached when the full sinuous form develops constrictions along its length with the subsequent breaking off of unit lengths which curve round to give the doughnut forms found in huge aggregates (Fig. 14).

This type of morphogenesis is by no means unknown in the virus world. For example, apart from the step giving rise to the torus, the means of assembly described here would satisfy the rhabdoviruses as well as the Marburg agent (BRADISH and KIRKHAM, 1966; NAKAI and HOWATSON, 1968). However, the dimensions of the Marburg agent differ considerably from those of the rhabdovirus group as the diagrammatic representation in Fig. 21 shows.

The other structure similar to the Marburg agent is that found in association with leptospires (ALMEIDA et al., 1969; PETERS and MÜLLER, 1969b). It now becomes clear that it is the empty sinuous form of the Marburg agent to which the leptospiral structures show a resemblance. This also is shown diagrammatically in Fig. 21 and it is of interest that the periodicity found in the Marburg agent, the leptospire-associated structure and vesicular stomatitis virus should all be 45 Å, suggesting that this may be a universal basic arrangement of lipid and protein, which always presents the same periodicity. The fact that the leptospiral structures resemble the *empty* Marburg particles makes the connection between the two less significant, for, as suggested, the 45 Å periodicity appears to be a general and non-specific biological phenomenon, while the Marburg internal helix which is absent from the leptospiral structures, probably contains the genetic material, and hence the specificity, of this agent. However, it would be equally wrong at the moment to disregard the similarity between these two structures, since in our present studies we found several examples of the Marburg agent in which the outer (presumably lipoprotein) membrane was spirally arranged, a feature of leptospires themselves, and also found on examples of the associated structures (Figs. 6, 7, 8).

While it is possible to carry out more detailed studies on the agent as present in tissue culture preparations and also in animal systems (the same structures as have been described here for tissue culture grown virus have also been seen in infected guinea-pig liver (ZLOTNIK and ALMEIDA, unpublished observations), it is important to establish whether these are derived from the agent present in the original monkeys. Immune electron microscopy, a technique used initially to characterise antibody molecules, has since been found equally valuable for characterising antigens, or the structures on which the antigens are located, when a specific antiserum is available (ALMEIDA and WATERSON, 1969). All the forms of the agent described were coated with antibody when antisera from human, guinea-pig and hamster sources were used (Figs. 18—20), but remained unaltered with control sera. This makes it virtually certain that the particles described in the literature are particles of the agent responsible for the outbreak at Marburg (HENNESSEN et al., 1968).

Finally there remains the question as to the nature of this agent, and where it fits into a classification of micro-organisms. Biologically, the Marburg agent seems to satisfy most of the criteria for a virus. It is cell-dependent for its multiplication, and would seem to possess only one nucleic acid (RNA), as it is capable of multiplication in the presence of Actinomycin D (KISSLING et al., 1968). In addition, the helical symmetry of the particles is an attribute of a great many viruses. However, the morphology that we have described here is not typical of any known virus, size alone making it sufficiently different from viruses such as vesicular stomatitis to exclude it from the rhabdovirus group. In addition to this size difference there is also the fact that the mature form of the virus would seem to be represented by the large aggregates of predominantly doughnut forms (Fig. 17). In view of this we would like to conclude by suggesting that we borrow the mathematical term 'torus', which may be defined as "a solid ring of circular section", and make the Marburg agent the founder member of a group of micro-organisms known as the 'toroviruses'.

Acknowledgements

We are grateful to Dr. D. Peters for a full and cordial exchange of information on the Marburg agent during these investigations. One of us (J. D. A.) is supported by a grant from the Medical Research Council. We acknowledge the support of the National Fund for Research into Crippling Diseases (Action for the Crippled Child). We are indebted to Miss Mary Gibbs for technical assistance.

References

Almeida, J. D., Waterson, A. P.: The morphology of virus-antibody interactions. Advances in Virus Res., **15,** 307 (1969).

Almeida, J. D., Waterson, A. P., Berry, D. M., Turner, L. H.: Structures associated with leptospires possibly relevant to the Marburg agent. Lancet **(1969) i,** 235—237.

Bradish, C. J., Kirkham, J. B.: The morphology of vesicular stomatitis virus (Indiana C) derived from chick embryos or cultures of BHK 21/13 cells. J. gen. Microbiol. **44,** 359—371 (1966).

Hennessen, W., Bonin, O., Mauler, R.: Zur Epidemiologie der Erkrankung von Menschen durch Affen. Dtsch. med. Wschr. **93,** 582—589 (1968).

Kissling, R. E., Robinson, R. Q., Murphy, F. A., Whitfield, S. G.: Agent of disease contracted from green monkeys. Science **160,** 888—890 (1968).

Kunz, C., Hofmann, H., Kovac, W., Stockinger, L.: Biologische und morphologische Charakteristika des Virus des in Deutschland aufgetretenen „Haemorrhagischen Fiebers". Wien. klin. Wschr. **80,** 161—162 (1968).

Martini, G. A.: Vervet monkey disease: human cases. Trans. Roy. Soc. trop. Med. Hyg. **63,** in press (1969).

May, G., Knothe, H., Hülser, D., Herzberg, K.: Elektronenmikroskopische Befunde bei einer Affenseuche (*Cercopithecus aethiops*). Zbl. Bakt. **(1968) I** (Orig.) **207,** 145—151.

Nakai, T., Howatson, A. F.: The fine structure of vesicular stomatitis virus. Virology, **35,** 268—281 (1968).

Peters, D., Müller, G.: Die elektronenmikroskopische Erkennung und Charakterisierung des Marburger Erregers. Dtsch. Ärzteblatt **65,** 1827—1834 (1968).

Peters, D., Müller, G.: Zur Morphologie und Entstehung des Marburg-Virus. Zbl. Bakt. I (Ref) **215,** 545 (1969a).

Peters, D., Müller, G.: The Marburg agent and structures associated with *leptospira*. Lancet **(1969b) i,** 923—925.

Shu, H. L., Siegert, R., Slenczka, W.: Zur Pathogenese und Epidemiologie der Marburg-Virus-Infektion. Dtsch. med. Wschr. **93,** 2163—2165 (tr. German med. Monthly (1969) **14,** 7—10) (1968).

Siegert, R., Shu, H. L., Slenczka, W., Peters, D., Müller, G.: Zur Ätiologie einer unbekannten, von Affen ausgegangenen menschlichen Infektionskrankheit. Dtsch. med. Wschr. **92,** 2341—2344, 2351—2352 (1967) (tr. German med. Monthly (1968) **13,** 1—3).

Waterson, A. P.: Measles virus. Arch. ges. Virusforsch. **16,** 57—80 (1965).

Waterson, A. P.: *Introduction to Animal Virology.* 2nd. Ed. Cambridge, 1968, 22—28.

Zlotnik, I., Simpson, D. I. H., Howard, D. M. R.: Structure of the vervet-monkey-disease agent. Lancet **(1968) ii,** 26—28.

Absence of Serological Relationship
Between the Marburg Virus and Some Arboviruses

Dr. J. Casals

It can be stated with hardly a doubt that the serological characterization of the Marburg agent is as complete as that of any infectious agent that has been investigated in recent years. The result of the investigation can be stated simply: Marburg virus is an agent *sui generis*, serologically distinct and unrelated to any of perhaps close to 200 microorganisms with which it has been compared. In a way this could end the present paper, particularly since a sizable part of the results directed towards establishing the antigenic characteristics of the agent have already been reported in publications from several laboratories.

There are some reasons, however, that justify the present communication:

a) An attempt to bring together all serological examinations carried out, with particular stress on viral studies and on the work done at Yale Arbovirus Research Unit-Rockefeller Laboratories.

b) A description of the phases through which the emphasis on the characterization of this agent has gone, from the very early days of the onset of the outbreak in August, 1967, to the present day.

c) A description of recent work, unreported, comparing the Marburg agent with other viruses, some of which bear certain similarities to it, for example in the type of illness caused by them.

a) Summary of Serological Studies

Other participants in this Symposium have or will report on studies dealing with the comparison of the Marburg agent with bacterial microorganisms, which early set this agent apart from a number of bacteriae (Schizomycetes), including leptospirae and rickettsiae. We will report here only on virological studies, mainly by the staff of YARU; but also, for the sake of completeness, I will mention studies done elsewhere, especially when done with viruses not available in our laboratory.

Owing to the extreme danger at first attached to work with the Marburg agent and to the lack in our organization of isolation facilities comparable to those of other laboratories, added to the fact that attempts to isolate and, later, propagate the agent were already under way in other laboratories, the decision was made that no work connected with isolation or propagation of the agent would be done at YARU. Consequently the work reported here consists almost exclusively of examination of the reaction between sera from patients and guinea pigs and viral antigens, excluding the Marburg antigen; in some recent tests, an inactivated Marburg virus antigen has been tested against various immune sera.

The sera and antigen pertaining to the Marburg infection received and, for the most part, studied at YARU are listed in Table 1.

Table 1. *Materials connected with Marburg virus infection or inoculation used in tests at YARU*

Sera

Human, supplied by Prof. R. SIEGERT and Dr. F. LEHMANN-GRUBE

Patient	Onset	Date of Sample
DIE	Aug. 14	Aug. 30, Sept. 9
FRI	Aug. 8	Aug. 21
HAN	Aug. 16	Sept. 9
HIL	Aug. 14	Aug. 30, Sept. 9
KLI	Aug. 14	Aug. 19, Aug. 24, Sept. 9
KRA	Aug. 14	Aug. 24, Sept. 9
MUE	Aug. 16	Sept. 9

Guinea pig
 Supplied by Drs. C. E. G. SMITH and D. I. H. SIMPSON, M.R.E.

Nos. 3699, 3700, 3701, and 3702.
 Bled 19 days after inoculation of infectious material, while convalescing; homologous complement-fixing titers, 1 : 32 — 1 : 64.

Nos. 1, 2, 3, and 4.
 Two immunized and two hyperimmunized with the established Marburg agent; CF titers, 1 : 128.
 Supplied by Dr. F. A. MURPHY, N.C.D.C.

No. 15, immunized with the Marburg agent; CF titer, 1 : 128.

Antigens
 Supplied by Dr. F. A. MURPHY, N.C.D.C.
 AGMA (Marburg), guinea pig liver, inactivated by bpl; CF titer, 1 : 128.
 Control for above, normal guinea pig liver.

In Table 2 are listed the arbovirus antigens and a few non-arbovirus ones employed in the tests. Some of the viruses that appear in the ungrouped class have now been placed in antigenic groups; but this is immaterial for the present report. One hundred and thirteen viruses are listed; however, considering group relationships and the manner in which the tests were conducted, it is fair to assume that these viruses represent an additional 50 arboviruses that belong to groups and are closely related by CF or HI to some of the listed agents.

In general outline the tests with sera listed in Table 1, and with antigens in Table 2 were done as follows:

1. Hemagglutination-inhibition test. The following sera were tested: from 6 patients bled on September 9, 1967, 24—26 days from onset of their illness and from 2 of the same patients bled on August 30, or 16 days from onset. Also were tested sera from 4 guinea pigs, convalescing 19 days from inoculation. While at the time that we carried out the tests the immune state of these sera was unknown owing to lack of an antigen for the Marburg agent, it was subsequently determined by Drs. SMITH and SIMPSON that CF antibodies were present in the sera. Hence, the negative results obtained at YARU could be considered valid.

Table 2. *Arbovirus antigens used in tests with sera from Marburg infection patients and from guinea pigs infected or immunized with the virus*

Group	Viruses
A	Aura[1], Chikungunya, EEE[1], Mayaro[1], Middleburg, Ndumu, O'nyong-nyong, Semliki, Sindbis, VEE[1], WEE.
B	Banzi, Bussuquara[1], dengues 1, 2, 3, and 4, IbAr8646, Ilheus[1], JBE[2], KFD, Langat[2], Louping ill, Ntaya, OHF, Powassan, RSSE, Spondweni, SLE[1], Tembusu[2], Uganda S, Wesselsbron, West Nile, yellow fever, Zika.
Bunyamwera	Batai, Bunyamwera, Cache Valley[1], Germiston, Ilesha.
Bwamba	Bwamba
C	Caraparu, Marituba[1], Oriboca
California	Tahyna, Lumbo
Guama	Catu
Simbu	Akabane, Ingwavuma, Sathuprei, Simbu, Yaba 7
Bakau	Ketapang
Dugbe	Dugbe, Ganjam
EHD	EHD-New Jersey
Kaisodi	Kaisodi, Lanjan
Kemerovo	Chenuda, Kemerovo
Koongol	Koongol, Wongal
Mossuril	Mossuril
Nyando	Nyando
Phlebotomus	Neapolitan, Sicilian
Piry	Chandipura, Piry
Quaranfil	Johnston Atoll, Quaranfil
Tacaribe	Amapari, Junin, Machupo[1], Pichinde, Tacaribe, Tamiami, Maru 10411[1], Maru 10886[1], Maru 12056[1]
Turlock	Umbre
VSV	Cocal, VSV-Ind, VSV-NJ
Ungrouped	Bhanja, CTF, Congo, Flanders, Hart Park, IbAr2709, IG 5139, IG 5287, IG 7481, IG 633970, Kern Canyon, Kowanjama, Lagos bat, Mapputta, Mt. Elgon bat, Nyamanini, Ogunpa, Pinn., Rift valley fever[3], SA An 4511, SA Ar 136, Tataguine, Thogoto, TR 5843, Uukuniemi, Wad Medani, Wanowrie, Witwatersrand, Yaba 1, and YM31.
Non-Arbovirus	Herpes, LCM, NDV, poxvirus, rabies, mumps[1], SHF[4]

[1] Drs. K. M. JOHNSON and P. A. WEBB.
[2] Drs. C. E. G. SMITH and D. I. H. SIMPSON.
[3] Dr. R. COSTLOW.
[4] Drs. A. SHELOKOV and N. TAURASO.

The above sera were tested in 2-fold dilutions beginning at 1 : 10 against 8 units of antigen; the method was practically the same at the MRE (Drs. SMITH and SIMPSON) as at YARU. The antigens used were: Group A, Chikungunya, O'nyong-nyong, Semliki, and Sindbis; Group B, dengue 1 and 2, West Nile, Spondweni, Banzi, Wesselsbron, Zika, yellow fever, Ntaya, RSSE, IbAn8646, JBE, Tembusu, Langat, and Louping ill; others, Bunyamwera, Germiston, Ilesha, Bwamba, Tahyna, and IbAn 2709.

2. Completement-fixation test. The sera from the 7 patients listed, including one from the first person to come down in the outbreak, were tested at YARU; the same specimens, or others close in time to them, were subsequently tested at

MRE, once they had prepared an antigen, and found to have CF antibody titers between 1 : 16 and 1 : 64, as stated above. The general procedure followed was to test each serum in 4 two-fold dilutions, beginning at either 1 : 4 or 1 : 8, against each antigen in dilutions 1 : 4, 1 : 8, and 1 : 16, which represented from 4 to 64 or more antigenic units. As far as it has been ascertained, the same procedure was followed at Middle America Research Unit, by Drs. Johnson and Webb.

The guinea pig sera were, in addition, tested against selected antigens, as described later.

The result of this considerable number of tests was invariably negative, with the exception of that given by sera from patients HIL which reacted at dilutions 1 : 4 or 1 : 8 with antigens for herpes and RSSE and DIE which reacted with a titer of 1 : 4 or 1 : 8 with herpes; these antibodies were considered unrelated to the current disease.

The conclusion that we reached and reported at the time, October 4, 1967, was that while negative results are not as definitive as positive evidence, it was our opinion that it was most unlikely that any of the viruses used in our survey was responsible for the illness of the 6 patients; this conclusion is still valid today.

b) Emphasis on Possible Particular Etiologies of the Marburg Infection

In the early days following the onset of the outbreak, the association between human illness and blood or tissues from African monkeys led to consider the possibility of an etiological agent originating in equatorial Africa; observation in postmortem specimens of hepatic lesions or formations reminiscent of Councilman bodies aroused the suspicion that the disease might be yellow fever, with humans infected through an aerosol. Rift valley fever was also mentioned in this connection.

By the 30th of August, 1967, yellow fever was no longer considered (at that time, tests at YARU with the first serum available, FRI, showed lack of positive reaction with yellow fever and other Group B antigens). The definite hemorrhagic manifestations both clinical and pathological, led the attending physicians to entertain the possibility of an hemorrhagic fever, probably caused by an arbovirus, since leptospiral infection did not appear to be at work.

The early reported results on the possible arboviral etiology of this hemorrhagic fever were somewhat conflicting. A laboratory reported the presence of Group B antibodies (against West Nile and Wesselsbron) "compatible with a Group B infection", in one of the patients. Another reported antibodies against Omsk hemorrhagic fever in 5 patients, with an increase in titer in one of them, leading to "thinking of the possibility that the responsible agent belongs in the TBE subgroup, closer to OHF than to the others". Another laboratory, somewhat later, observed questionable antibodies in one of the patients against some of the newly discovered viruses of the Tacaribe Group.

While there may have been clinical and pathological similarities between the Marburg virus infection and the viral hemorrhagic fevers—both of Group B and outside—the serological evidence gathered at YARU at the time did not support the theory of a serological connection between the disease and any of the Tacaribe

Group agents available or with the hemorrhagic fevers of the TBE complex; a summary of the results is shown in Table 3.

Table 3. *Complement-fixation test*
Human convalescent and guinea pig immune sera, Marburg virus
Tacaribe group and 2 group B antigens

Antigen Reciprocal of Titer		Serum												
		Man, Aug. 30, 1967			Man, Sept. 9, 1967						Guinea Pig			
		DIE	HIL	DRE	DIE	HIL	KRA	MUE	HAN	KLI	99	00	01	02
Amapari	256	0	0				0, 0		0		0	0	0	0
Junin	256	0	0, 0	0	0	0	0, 0	0	0	0, 0	0	0	0	0
Machupo			0	0			0		0					
Pichinde	16	0	0								0	0	0	0
Tacaribe	256	0	0, 0	0			0, 0			0, 0	0	0	0	0
Tamiami	256	0	0, 0	0			0, 0			0, 0	0	0	0	0
Maru 12056			0	0			0		0					
Maru 10411			0	0			0		0					
Maru 10886			8 ?				16 ?		16 ?					
KFD	256	0	0		0	0	0	0	0	0				
Omsk HF	512	0	0		0	0	0	0	0	0				
Marburg			32	32	32		16–64	16–64				32–64	32–64	32–64	32–64

Reciprocal of serum titers; 0, no fixation at dilution 1 : 4 or 1 : 8.
When duplicate results appear one was by Dr. K. M. Johnson, who also performed all the tests with serum DRE and with antigens Machupo and Marus.

As can be seen in the table, at no time was there in our hands a positive reaction between human or guinea pig sera, known subsequently to be positive by CF against the Marburg agent, and any of the Tacaribe Group agents or the antigens from the TBE complex. The positive reaction given by 3 sera and antigen MARU 10886, first reported by Drs. Johnson and Webb, were subsequently considered by the same workers to have been non-specific.

In October, 1967, Professor Siegert first described the morphology, now familiar, of the Marburg agent, later confirmed at the NCDC, Atlanta. Except for its dimensions, Marburg virus resembled, in its inner structure and configuration, the viruses of Vesicular Stomatitis Group. Based on its morphology, the working hypothesis was made that possibly Marburg virus was a rhabdovirus and that it might be serologically related to the VSV group. Among our early tests (see Table 4), no positive reaction was observed between the sera from the 6 patients and 2 antigens (VSV-Indiana and Chandipura). The tests were later extended to include VSV-New Jersey and Hart Park antigens with the serum of a patient; nor was there any positive reaction between 3 guinea pig sera from MRE and 5 antigens of the group. Recently, and as the group has expanded to include other agents, a comprehensive study has been done by Dr. R. E. Shope, YARU, whose results we quote with his kind permission. As Table 4 shows, no reaction has occurred between the serum from guinea pig Nr. 15 and any of the antigens; nor between antisera for each virus of the VSV group and Marburg antigen. The conclusion at this time is that while Marburg virus may be a rhabdovirus, there is no evidence that it is a member of the VSV antigenic group of arboviruses.

Table 4. *Complement — fixation test*
Human convalescent and guinea pig immune sera, Marburg virus
Rhabdovirus antigens

Antigen	Serum									
	Man, Sept. 9, 1967						Guinea Pig			
	DIE	HIL	KRA	MUE	HAN	KLI	2	3	4	15
VSV-NJ			0				0	0	0	0
VSV-Ind	0	0	0	0	0	0	0	0	0	0
Cocal							0	0	0	0
Piry										0
Chandipura	0	0	0	0	0	0				0
Hart Park			0				0	0	0	0
Flanders										0
Rabies							0	0	0	0
Mt. Elgon bat										0
Kern Canyon										0
Marburg	32*		16–64*	16–64*			64	64–128*	64–128*	128

Reciprocal of serum titers; 0, no fixation at dilution 1 : 4.
Guinea pig sera No. 2, 3, and 4 supplied by Dr. D. I. H. SIMPSON; No. 15 by Dr. F. A. MURPHY.
   * Titers supplied by Drs. D. I. H. SIMPSON and C. E. G. SMITH;
Results with guinea pig No. 15 supplied by Dr. R. E. SHOPE; in a reciprocal test with sera for all above listed viruses against Marburg antigen, he obtained only negative reactions.

c) Recent Studies Involving Marburg Virus

Certain developments have occurred in recent months that have re-awakened our interest in continuing an examination of relationships between the Marburg agent and other viruses. CHUMAKOV and associates have succeeded lately in isolating in mice the causative virus of Crimean hemorrhagic fever and we have shown that this virus is similar, if not identical, antigenically to Congo virus; the latter has been known to be present in equatorial Africa from Uganda to Nigeria. Still more recently a hemorrhagic disease has been observed by FRAME and associates, which at this writing has affected 3 American medical missionary nurses, stationed in NE Nigeria; of these, 2 died and the third recovered after a long and serious illness. An additional as yet unreported illness (with virus isolation) has occurred in a staff member (J. C.) of YARU. From the early sera from all 3, Dr. BUCKLEY, YARU, has isolated a virus by inoculation of African green monkey kidney cells cultures (Vero cells). This agent is currently under investigation; serologically, it has been found distantly related to LCM and Tacaribe Group viruses of all agents with which it has so far been compared. We have succeeded in developing a CF system with this agent, called here Pinn., and studies carried out with it, Marburg and other systems are summarized in Table 5.

Table 5. *Complement-fixation test*
Study of relationships among viral hemorrhagic fevers

Antigen, titer	Serum Marburg			Con	Pinn.		SHF	NSD	RVF	EHD	KHF	G.B.	G.TAC
	Gp2	Gp15	Man	Mou.	Man	Mou.	Monk.	Mou.	Sheep	Mou.	Man	Mou.	Mou.
Marburg			32–										
1 : 128	64	128	64	0	0		0	0	0	0	0	0	0
Congo				128–									
1 : 1024		0	0	256	0	0	0	0	0	0		0	0
Pinn													
1 : 16	0			0	128	128	0	0	0	0	0	0	0
SHF													
1 : 256			0***	0	0		256	0	0	0	0***	0***	0***
NSD								128*					
RVF									256**				
EHD-NJ													
1 : 16		0	0		0	0				16			
Group B													
1 :128–512	0	0	0	0			0	0	0	0	0	Pos	0
Tacaribe g.													
1 : 32–512			0	0	0		0			0	0	0	Pos

SHF, simian hemorrhagic fever; NSD, Nairobi sheep disease; RVF, Rift Valley fever; EHD-NJ, epizootic hemorrhagic deer disease, New Jersey type; KHF, Korean hemorrhagic fever.

Group B antigens include: Omsk HF, KFD and dengues 1 — 4; Tacaribe group include: Amapari, Junin, Machupo, Pichinde, Tacaribe, and Tamiami.

Reciprocal of serum titers; 0, no fixation at 1 : 4 or 1 : 8.

   * Titer given by Dr. R. West, E.A.V.R.I.
  ** Titer given by Dr. Richard Costlow.
 *** Results supplied by Drs. N. Tauraso and A. Shelokov.

The observations with the viruses listed in Table 5 are, at the moment, still incomplete. With some agents, Nairobi sheep disease and Rift Valley Fever, no antigens are available owing to safety restrictions; with another system, Korean hemorrhagic fever, no virus has as yet been propagated in laboratory experimental hosts, therefore, sera from survivors can only be assumed to have antibodies. The results with Simian hemorrhagic fever will be discussed more extensively by Drs. Tauraso and Shelokov later in the Symposium. The net result of these investigations is that the 10 types of hemorrhagic fevers shown in the table are antigenically unrelated.

Conclusion. The result of an extensive search for antigenic relationships between the Marburg virus and a large number of arboviruses, and a few viruses not of that set, can be summarized with the sentence with which I opened this paper. Marburg virus is an antigenically distinct virus, unrelated to any with which it has been so far compared.

Biological Properties of the Marburg Virus

W. Slenczka and G. Wolff

The Marburg virus appears to have a wide host range. Growth of virus was reported in several species of monkeys [2, 3, 4, 14, 15], in guinea pigs [7, 11, 12, 13, 18], and suckling mice [7], in primary and continuous cell lines of various origin [1, 2, 5, 6, 7, 9, 10, 11, 13, 16, 18, 19, 20], and in insects [8]. However, in adult mice [7] and in the fetal organs of chick-embryos [18, 11] no multiplication of virus could be detected. Susceptibility to infection with Marburg virus varies considerably from host to host. In most systems the pathogenicity of the virus for a new host seems to increase gradually during several passages.

One of the most noteworthy observations with human infections was the excretion of virus with the semen, which we detected in one case 12 weeks after the onset of the disease. Since the patient's wife had contracted the disease from her husband one week previously, virus transmission by sexual intercourse is highly probable in this case. Theoretically, a vertical spread of the virus from the infected parent to the progeny could be a consequence of virus excretion with the semen. Carrier state and vertical spread would be of considerable importance for biological and epidemiological reasons. We have therefore tried to establish a persistent infection in various hosts and to study the possibility that the virus is transmitted to the progeny.

In the organs of guinea pigs which had survived a low dose of the Marburg virus from a low passage number, we have never detected viral antigen by immunofluorescence after the febrile phase had ended. Two pregnant guinea pigs were infected some days before giving birth. Both animals died along with the fetuses. In both cases numerous viral inclusions in the organs of the mother animals were found. However, in the placentae and in the fetal organs (a total of five) neither infectious virus nor viral antigen could be demonstrated. This suggests that transmission of virus via the placenta does not occur in guinea pigs during this stage of pregnancy. Nevertheless, virus transmission via germinal cells is still a posibility in spite of this negative result.

Adult mice do not show signs of disease after infection with Marburg virus. According to Kunz and coworkers [7] the virus multiplies in newborn mice. We have inoculated adult albino mice via the intraperitoneal and intracerebral routes. No viral inclusions were found in the organs of these animals seven and 14 days after infection. Sera of ten of these animals were collected six weeks after the inoculation. They were pooled and tested for the presence of complement-fixing antibodies; the titre was 1: 32.

A lasting persistence of virus was observed in mice infected neonatally when less than 24 h. old. A total of 140 new-born mice were inoculated intracerebrally with Marburg virus of the first, second, and third passages in new-born mice. The deathrate of these animals during the first week of life was about 30%, which was probably due to the trauma of the intracerebral inoculation. The brains of those animals, which were examined in the course of the first two weeks after infection, contained foci of cells with typical intracytoplasmic inclusions. In livers and spleens only small amounts of antigen were encountered. The organs of these animals were infectious. In the following weeks some of these animals developed normally and never showed signs of disease. Blood of these mice was collected six weeks after infection. It was not infectious for guinea pigs. Titres of CF-antibodies ranged between 1 : 8 and 1 : 32. Four animals in this group were killed six weeks after infection and their organs were tested for the presence of antigen by immunofluorescence; occasional cells with typical intracytoplasmic inclusions were found in two of these animals.

About 30% of the mice which were inoculated neonatally and had survived the first weeks were slightly retarded in their development. About three weeks after the inoculation these animals looked sick; they were prostrated and sat motionless in their cages with ruffled fur. All died within three to four weeks after infection. At this time the number of cells with cytoplasmic inclusions was significantly greater in the organs of the sick mice than in the organs of mice without signs of illness. Surprisingly, histopathological examination of the livers, spleens, and brains of these mice—kindly performed by Dr. Korb and Dr. Solcher—did not reveal pathological alterations. Thus it is doubtful whether the infection with Marburg virus has actually been the cause of death in these animals.

The answer to the question whether the persistence of virus in the surviving mice is permanent, and whether a vertical spread is possible must await examination of the second generation of these mice.

In Vero cell cultures a chronic infection with Marburg virus was established without difficulty. After infection of these cells a CPE develops only with very high infectious doses. In order to establish a chronically infected cell line, Vero cells were infected with a low dose of Marburg virus, adapted by six serial passages. At the fourth day after infection the cells were dispersed by treatment with versene, and subcultures were prepared in the usual way. Viral antigen was found in more than 90% of the first and the following subcultures; few cells seemed to be free of virus. No CPE was observed in these chronically infected cells. As compared with non-infected control cultures, the growth rate of the chronically infected cells does not seem to be impaired. In the fifth subculture cell-associated and extracellular virus from the fourth day after preparation were harvested and titrated separately. The titres, which were determined by reading the endpoints in cultures after staining with fluorescent antiserum, were 10^5 and 10^6 ID_{50}/ml for cell-associated and extracellular virus, respectively. Similar virus yields were found in Vero cells after primary infection with Marburg virus. The mechanism of this virus carrier state is not yet clear. Antiserum is not required for its establishment and maintenance. It is not known whether interferon or partial susceptibility of the cells is responsible for maintenance of the carrier state.

Because of their very high antigen content we have employed Marburg virus carrier cells for production of CF antigen. Crude antigens prepared from spleens and livers of infected guinea pigs, which have been used by others in the complement-fixation test, did not give reliable results in our hands. Normal human sera often exhibited high non-specific titres against antigens from the livers of infected guinea pigs, as well as against control antigens from livers of normal guinea pigs. On the other hand, human convalescent sera reacted with the control antigens often to the same extent as with virus antigen. In our experience crude antigens from guinea pig livers are so unreliable that in doubtful cases no decision can be made as to whether a titre is specific or not. The antigen which we have prepared from carrier cells did not react non-specifically. In one case, in which the physicians suspected that a woman might have contracted the disease from her husband more than one year after his infection, the CF-test with her serum revealed titres of 1 : 8 with both infected guinea pig liver and control antigens. When the test was repeated using the cell culture antigen, no reactivity with either infectious antigen or with control antigen was detected. Thus an infection with Marburg virus could be ruled out with confidence.

It may be said that, in our experience with Vero cells, neutralization tests in which CPE is taken as the indicator may very often lead to falsely positive results. In these cells CPE of Marburg virus is markedly inhibited when normal human serum at concentrations of 5 or 10% is included in the medium, although growth of virus is not inhibited. A more reliable neutralization test may be carried out in vitro when Vero cells with cytoplasmic inclusions are counted, and their reduction taken as a measure for neutralization [16, 17].

Results of Complement-Fixation Tests Carried Out with Antigens from Guinea Pig Liver and from Vero Cells.

Sera	Days	no antigen	guinea pig liver antigen normal	guinea pig liver antigen infectious	Vero cell antigen normal	Vero cell antigen infectious
1 Kr.	12	—	1 : 32	1 : 64	—	1 : 16
2 Kr.	33	—	1 : 32	1 : 64	—	1 : 32
3 Kr.	45	—	1 : 16	1 : 32	—	1 : 16
4 Kr.	617	—	1 : 32	1 : 64	—	1 : 16
5 Ul.	6	—	—	—	—	—
6 Ul.	20	—	—	1 : 32	—	1 : 32
7 Ul.	29	—	1 : 16	1 : 16	—	1 : 16
8 Ul.	42	—	1 : 8	1 : 16	—	1 : 32
9 Ot.	4	—	—	—	—	—
10 Ot.	16	—	1 : 2	1 : 4	—	1 : 32
11 Ot.	38	—	1 : 8	1 : 16	—	1 : 16
12 Ot.	122	—	—	1 : 4	—	1 : 16
13 Hi.	122	—	1 : 4	1 : 32	—	1 : 16
14 Ul., K.	—	—	1 : 8	1 : 8	—	—
15 We.	—	—	1 : 16	1 : 32	—	—
16 Ko.	—	—	1 : 16	1 : 16	—	—
17 Wo.	—	—	—	—	—	—
18 G. P. 113	67	—	n. d.	n. d.	—	1 : 128
19 G. P. 159	26	—	—	1 : 40	n. d.	n. d.

No. 1—13: Sera from four human patients.
No. 14: Serum from a doubtful case in which infection with Marburg virus had been suspected.
No. 15—17: Sera from three normal persons.
No. 18—19: Anti-Marburg virus hyperimmune sera from two guinea pigs.

In closing I wish to report a preliminary observation, which, however, requires confirmation. Prof. Peters has stressed the morphological similarities which exist between Marburg and rabies viruses. We have done some experiments in search of an antigenic relationship between these viruses.

No cross immunity exists; guinea pigs which were immunized with rabies virus were not protected against Marburg virus, and vice versa.

CF tests with three anti-rabies hyperimmune sera from guinea pigs and antigen from Marburg virus gave negative results. The CF titres of these sera with rabies antigen have not yet been determined.

Using the direct method of immunofluorescence, Negri bodies in rabies-infected BHK cells stained with conjugates of Marburg virus antisera from three guinea pigs. Conjugates of sera from patients convalescing from Marburg virus disease did not stain Negri bodies. Using a commercial antirabies conjugate, we were unable to stain cytoplasmic inclusions in Vero cells infected with Marburg virus. I am aware that no final statement regarding antigenic relationship may be based on the results of direct immunofluorescence only. Further investigations are needed to elucidate the nature and significance of these findings.

References

1. Carter, G. B., Bright, W. F.: Lancet **1968 II,** 913—914.
2. Haas, R., Maass, G., Müller, J., Oehlert, W.: Z. med. Mikrobiol. und Immunol. **154,** 210—220 (1968).
3. Haas, R., Maass, G., Oehlert, W.: Med. Klinik **63,** 1359—1363 (1968).
4. Haas, R., Maass, G., Oehlert, W.: Primates in Med. **3,** 135—137 (1968).
5. Hofmann, H., Kunz, Ch.: Zbl. Bakt. I Orig. **208,** 344—347 (1968).
6. Kissling, R. E., Robinson, R. Q., Murphy, F. A., Whitfield, S. G.: Science **160,** 888—890 (1968).
7. Kunz, Ch., Hofmann, H., Kovac, W., Stockinger, L.: Wien. klin. Wschr. **80,** 161—162 (1968).
8. Kunz, Ch., Hofmann, H., Aspöck, H.: Zbl. Bakt. I Orig. **208,** 347—349 (1968).
9. Maass, G. et al.: Morbidity and Mortality: U.S. Department of Health. Follow-up obscure disease related to African Monkeys. Weekly Rep. **17,** 223 (1968).
10. Malherbe, H., Strickland-Cholmley, M.: Lancet **1968 II,** 1434.
11. May, G., Knothe, H.: Dtsch. med. Wschr. **93,** 620—622 (1968).
12. Siegert, R., Shu, H. L., Slenczka, W., Peters, D., Müller, G.: Dtsch. med. Wschr. **92,** 2341—2343 (1967).
13. Siegert, R., Shu, H. L., Slenczka, W.: Dtsch. med. Wschr. **93,** 604—612 (1968).
14. Simpson, D. I. H., Zlotnik, I., Rutter, D. A.: Brit. J. Expt. Pathol. XLIX, 458—464 (1968).
15. Simpson, D. I. H., Bowen, E. T. W., Bright, W. F.: Lab. Anim. **2,** 75—81 (1968).
16. Slenczka, W.: 2. Arbeitstagung Dtsch. Ges. Hygiene und Mikrobiologie, Mainz, 7. und 8. Okt. 1968.
17. Slenczka, W.: 29th. Symp. of I.A.M.S.: Hazards of Handling Non-Human Primates, 8th to 11th April 1969, Brighton. Lab. Anim. Handb. 4, 143—147 (1969).
18. Smith, C. E. G., Simpson, D. I. H., Bowen, E. T. W., Zlotnik, I.: Lancet **1967 II,** 1119—1121.
19. Zlotnik, I., Simpson, D. I. H., Bright, W. F., Bowen, E. T. W., Batter-Hatton, D.: Brit. J. Expt. Pathol. XLIX, 311—314 (1968).
20. Zlotnik, I., Simpson, D. I. H.: Lancet **(1968) II,** 205.

Some Characteristics of the Marburg Virus

Ch. Kunz and H. Hofmann

In September 1967, we received some specimens from patients with Marburg virus disease for virus isolation purposes. With these specimens, kindly supplied by Professor Siegert, we conducted a series of studies the results of which will be summarized in this report.

Assuming that the agent could possibly be an arbovirus, we injected blood from a patient intracerebrally into newborn mice. During an observation period of 21 days the mice showed no signs of illness. However, guinea pigs injected with brain of mice sacrificed on the 5th day developed typical disease. The agent could be propagated in the mice in serial passages, which were carried out at 5-day intervals. Yet, even after 20 passages, no fatalities were observed and the mice only underwent a silent infection.

The identity of the agent isolated by us in mice, with the etiologic agent of the disease, was ascertained in the following manner:

1. Viral antigen could be stained in liver imprints of guinea pigs with fluorescein-labeled antisera from both patients and immunized guinea pigs.

2. In electron microscopic studies, done with thin-cuts of guinea pig liver and spleen, the typical rod-shaped viria were found and

3. from the infected liver of guinea pigs, a crude antigen was obtained which reacted in the CF test to a high titer with sera taken from convalescent patients.

In a second experiment we demonstrated that the agent is sensitive to 1% Na-desoxycholate which inactivates $10^{4.5} LD_{50}$ of the virus.

Further studies dealt with the question of whether or not the virus multiplies in artificially infected arthropods.

For investigating the ability of the Marburg virus to replicate in mosquitoes, we first employed *Aedes egypti*. The mosquitoes were injected intrathoracally with a guinea pig serum containing $10^7 LD_{50}$ of virus per ml. The infected *Aedes egypti* were kept in an incubator at 26 °C. A high humidity was maintained at all times. On the 11th day surviving mosquitoes were ground and suspended in PBS. The suspension was injected into guinea pigs and also again into *Aedes* for a second passage. Mosquitoes of the second passage were sacrificed 21 days after infection and tested for virus in guinea pigs. From both passages virus could be reisolated in guinea pigs and was readily identified as Marburg virus by means of fluorescent antibodies. The result of this experiment shows that the Marburg virus can be propagated in *Aedes egypti* mosquitoes.

By contrast, Marburg virus obviously failed to replicate after the artificial infection of *Anopheles maculipennis* mosquitoes which belong to the subgroup *Anophelinae*. In these mosquitoes, virus was detectable on the 2nd, 4th, and 8th days, respectively, after inoculation but not thereafter.

In another experiment propagation of the virus in ticks of the species *Ixodes ricinus* was attempted. Female ticks were inoculated by the anal route using, as in the previous tests with mosquitoes, a thin glass capillary and viraemic guinea pig serum. Also in this experiment, we found no evidence for multiplication of Marburg virus in this type of arthropod. No virus was demonstrable in ticks of the first virus passage harvested on the 15th day after infection and in ticks of the second passage which were ground and assayed for virus in guinea pigs on the 18th day.

Since all mosquito-borne arboviruses so far tested multiply in *Aedes egypti* after the intrathoracic infection, it seems to be a reasonable assumption that the Marburg virus can be transmitted in nature by mosquitoes. In view of the fact that the virus could not be propagated in *Anopheles maculipennis*, the vector probably is a mosquito of the subgroup *Culicinae* of which *Aedes egypti* is a member. We must, however, emphasize that our results do not exclude the possibility of transmissions of the virus without an arthropod as vector, as is the case with Vesicular Stomatitis virus.

Recently, we investigated the capacity of the Marburg virus of inducing the formation of interferon in baby mice. For this purpose, baby mice were infected intracerebrally with high doses of the virus. Brains were removed from some mice on the 5th, 8th, and 16th days, respectively, after infection and suspended in distilled water. The suspensions were dialysed against citrate buffer pH 2, phosphate buffer pH 7.5, and finally against distilled water. Eagle's medium of 5-fold concentration was added 1 : 4. Two-fold dilutions of this preparation were tested to prevent CPE of $100-300$ TCD_{50} of EMC virus in L cells. In this system no interferon-inducing ability of the Marburg virus was found.

Because of the morphologic similarity of the Marburg virus to the Rhabdovirus group, Vesicular Stomatitis virus (VSV), Hart Park, and Cocal viruses were also investigated. A high level of interferon was observed in the brains of baby mice after the infection with Hart Park virus and a low titer of interferon was found with Cocal and VS viruses.

Formation of interferon is, therefore, not a general characteristic of the Rhabdo group of viruses.

Finally, we tested the sensitivity of the Marburg virus to Poly I : C a double-stranded polyribonucleotide which is capable of inducing the formation of interferon, as demonstrated by several investigators. Recently, we could show that this compound is able to protect mice against fatal tick-borne encephalitis.

In this experiment, 27 baby hamsters weighing $12-15$ g were treated with 100 μg Poly I : C 18 and again 3 hours before intracerebral infection with Marburg virus. The virus used had undergone 3 passages in livers of hamsters. A dilution of 10^{-3} was used. As controls, another group of 27 hamsters was infected but not treated.

From the group treated with Poly I : C, 14 hamsters died and from the controls 13 animals succumbed the disease. The average survival time was 7.79 and 8.85 days. Thus, Poly I : C did not protect hamsters against infection with Marburg virus.

The results of our studies can be summarized as follows:

The virus replicates in the brains of baby mice without causing disease and is sensitive to desoxycholate.

Marburg virus could be propagated in artificially infected mosquitoes of the species *Aedes egypti* but not in *Anopheles maculipennis*. It also failed to replicate in *Ixodes ricinus* ticks. These findings suggest that the virus is possibly transmitted in nature by mosquitoes of the *Culicinae* group of which *Aedes egypti* is a member.

In our experience the virus is not capable of inducing the formation of interferon nor does it appear to be sensitive to this viral inhibitor. This is indicated by the failure of the interferon-inducing compound Poly I : C to protect hamsters against the infection with Marburg virus.

References

KUNZ, CH., HOFMANN, H., KOVAC, W., STOCKINGER, L.: Biologische und morphologische Charakteristika des Virus des in Deutschland aufgetretenen „Hämorrhagischen Fiebers". Wien. Klin. Wschr., **80**, 161—162 (1968).

HOFMANN, H., KUNZ, CH.: Das Verhalten des sogenannten „Marburg-Virus" in einigen Gewebekulturen. Zbl. Bakt., I, Orig., **208**, 344—347 (1968).

KUNZ, CH., HOFMANN, H., ASPÖCK, H.: Die Vermehrung des „Marburg-Virus" in Aedes aegypti. Zbl. Bakt., I, Orig., **208**, 347—349 (1968).

HOFMANN, H., KUNZ, CH.: Komplementbindende Antikörper nach Infektion mit dem „Marburg-Virus" (Rhabdo-Virus simiae) beim Menschen. Zbl. Bakt., I, Orig., **209**, 288—293 (1969).

HOFMANN, H., KUNZ, CH.: Interferonbildung im Gehirn weißer Säuglingsmäuse nach Infektion mit einigen Rhabdoviren. Zbl. Bakt., I, Orig., **211**, 5—9 (1969).

KUNZ, CH., HOFMANN, H.: Die Beeinflussung der experimentellen Frühsommer-Meningoenzephalitis (FSME)-Virusinfektion durch die Interferon-induzierende Substanz Poly I : C. Zbl. Bakt., I, Orig., **211**, 270—273 (1969).

HOFMANN, H., KUNZ, CH., ASPÖCK, H., RADDA, A.: Zur Ökologie des sogenannten „Marburg-Virus" (Rhabdovirus simiae). Zbl. Bakt., I, Orig., **212**, 168—173 (1969).

Cultivation of the Marburg Virus (Rhabdovirus simiae) in Cell Cultures

H. HOFMANN and CH. KUNZ

With 3 Figures

At the time when the causative agent of Marburg haemorrhagic fever was still unknown, attempts were made to isolate the agent from patients. For isolation purposes laboratory animals and many cell systems were employed but all initial trials in cell cultures were unsuccessful (SIEGERT et al., 1967; MAY and KNOTHE, 1968; SIEGERT et al., 1968). Later when the virus was isolated in guinea pigs (SIEGERT et al., 1967; SMITH et al., 1967; KUNZ et al., 1968; MAY and KNOTHE, 1968; KISSLING et al., 1968), many investigators tried to propagate Marburg virus in various cell cultures. Presently, much data are available which will be summarized in this report.

1. Propagation of Virus in Primary Cell Cultures (see Table 1)

Since the agent was found to replicate in man, monkey, and guinea pigs, cell cultures from these sources were initially employed.

Table 1. *Propagation of Marburg virus in primary cell cultures*

Primary cells	Virus replication	CPE	Virus titer	References
Cercopithecus kidney	+	−	n. d.*	SIEGERT et al., (1968)
Cercopithecus kidney	+	+	n. d.	HAAS et al., (1968)
Rhesus kidney	+	−	n. d.	HOFMANN and KUNZ, (1968)
Human amnion	+	+/−	n. d.	MAY et al., (1968)
Human leucocytes		−	n. d.	MAY and KNOTHE, (1968)
Chick embryo fibroblasts	+	−	10^1 g. p.**	HOFMANN and KUNZ, (1968)
Chick embryo fibroblasts	−			SMITH et al., (1967)
Chick embryo fibroblasts		−	n. d.	MAY and KNOTHE, (1968)
Guinea pig fibroblasts	+´	−	10^3 g. p.	HOFMANN and KUNZ, (1968)
Guinea pig kidney		−	n. d.	MAY and KNOTHE, (1968)

* n. d. = not done
** g. p. = assayed in guinea pigs

In primary cells of Cercopithecus kidney, the species with which the agent was imported to Germany, SIEGERT et al. (1968) found that the virus replicated without cytopathic effect (CPE). They also demonstrated the virus specific antigen by means of the immunofluorescent method. On the contrary HAAS et al. (1968)

observed gross CPE in primary Cercopithecus kidney cells. These cells were only less sensitive to the virus than subcutaneously infected Cercopithecus monkeys. In primary Rhesus monkey kidney cells, the virus replicated but without CPE (HOFMANN and KUNZ, 1968).

Primary cells of human origin also propagated the virus. MAY and KNOTHE (1968) could not observe CPE in human leucocytes but reported slight CPE in primary human amnion cells (MAY et al., 1968).

We could demonstrate replication of the virus in primary guinea pig embryo fibroblasts but no CPE was seen (HOFMANN and KUNZ, 1968).

In chick embryo cells, we found a very slight virus replication (HOFMANN and KUNZ, 1968), while other investigators did not observe propagation at all (SMITH et al., 1967; MAY and KNOTHE, 1968).

2. Propagation of Virus in Established Cell Lines (see Table 2).

Established cell lines were also investigated for propagation of Marburg virus. At first cells derived from monkeys were tested. Although we found that permanent Cercopithecus kidney cells (strain GMK-AH$_1$) produced high titers of virus— 1 ml of culture fluid contained 10^6 infective doses for guinea pigs—we could not demonstrate any CPE. KISSLING et al. (1968) were more successful; they observed CPE in their cultures in the second day p.i., which was in total about the 4th—5th day.

In the VERO cell line, the virus also propagated. No CPE was observed by SIEGERT et al. (1968) and only slight CPE by KISSLING et al. (1968).

From cells of Rhesus monkey origin, heart cells (strain CMH) allowed only slight virus growth (HOFMANN and KUNZ, 1968), while in kidney cells (strain LLC-MK$_2$) no virus replication was demonstrable (SMITH et al., 1967).

In the studies of SMITH et al. (1967) the Marburg virus propagated in L cells (mouse embryo cells) without CPE; we could not demonstrate any virus growth in those cells.

Heart cells derived from guinea pigs also allowed virus replication without CPE (KISSLING et al., 1968).

The first reports of Marburg virus-induced CPE in cell cultures were by ZLOTNIK et al. (1968). They had found that the virus was adaptable to BHK$_{21}$ cells. In the first passage typical inclusion bodies similar to those found in guinea pig liver, were seen in infected cells after 13 days and CPE appeared about the 23rd day. After a few passages, inclusion bodies as well as CPE appeared earlier. In our laboratory BHK$_{21}$ cultures showed only slight changes which appeared very late (HOFMANN and KUNZ, 1968). KISSLING et al. (1968) tested two strains of BHK$_{21}$. One, the WI 2 strain, behaved as the strain in our laboratory, but the other was highly susceptible. Cytopathic effect was observed about the 2nd—5th day after infection.

In contrast to 3 other teams of investigators (SIEGERT et al., 1968; SMITH et al., 1967; MAY and KNOTHE, 1968), we could propagate Marburg virus quite well in our HeLa strain. One ml of culture fluid contained 10^4 infective particles for guinea pigs, however we found no CPE (HOFMANN and KUNZ, 1968).

Other strains deriving from human sources were also tested for Marburg virus induced CPE. KISSLING et al. (1968) propagated the virus in foreskin fibroblasts.

Table 2. *Propagation of Marburg virus in established cell lines*

Cell line	Virus repli-cation	CPE	Virus titer	References
Cercopithecus kidney (GMK—AH$_1$)	+	—	10^6 g. p.**	HOFMANN and KUNZ (1968)
Cercopithecus kidney (GMK—AH$_1$)	+	+ 2nd — 5th	n. d.*	KISSLING et al., (1968)
VERO	+	—	n. d.	SIEGERT et al., (1968)
VERO	+·	+/—	n. d.	KISSLING et al., (1968)
Rhesus monkey heart (CMH)	+	—	10^1 g. p.	HOFMANN and KUNZ (1968)
Rhesus monkey kidney (LLC—MK$_2$)	—			SMITH et al., (1967)
L (mouse embryo)	—			HOFMANN and KUNZ (1968)
L (mouse embryo)	+	—	n. d.	SMITH et al., (1967)
Guinea pig heart	+	—	n. d.	KISSLING et al., (1968)
BHK$_{21}$	+	+/—	10^1 g. p.	HOFMANN and KUNZ (1968)
BHK$_{21}$	+	+ 7th — 23rd	n. d.	ZLOTNIK et al., (1968)
BHK$_{21}$ (W 12)	+	+/—	n. d.	KISSLING et al., (1968)
BHK$_{21}$ (CCL 10)	+	+ 2nd — 5th	$10^{6.5}$ t. c.***	KISSLING et al., (1968)
HeLa	+	—	10^4 g. p.	HOFMANN and KUNZ (1968)
HeLa	—			SIEGERT et al., (1968)
HeLa	—			SMITH et al., (1967)
HeLa		—	n. d.	MAY and KNOTHE, (1968)
Foreskin fibroblasts	+	+ but not in serial passages	n. d.	KISSLING et al., (1968)
U (Human Amnion)	+	—	n. d.	HOFMANN and KUNZ unpubl.
ELF (Embryonal human lung fibroblasts)	+	+ 3rd — 5th	10^6 g. p. 10^4 t. c.	HOFMANN and KUNZ unpubl.

   * n. d. = not done
  ** g. p. = assayed in guinea pigs
*** t. c. = assayed in tissue culture

In the first passage of virus CPE was demonstrable but could not be reproduced in serial passages. In U cells, a stable cell line of human amnion, the agent also replicated but without CPE (HOFMANN and KUNZ, unpublished).

In our laboratory we had previously tested many cell systems, but we were unable to detect a cell line in which Marburg virus propagates with CPE. Finally we came across the ELF (embryonal human lung fibroblasts) cell strain, in which CPE appears about the 3rd day and reaches its maximum about the 5th day after infection. Cytopathic effect begins in focal areas and consists of spindling and later on of clumping of cells. Finally the foci become confluent (see Figs. 1—3). It must be mentioned that, although changes are severe, they are never complete and eventually healthy cells may grow in and repair the lesions. Tissue culture

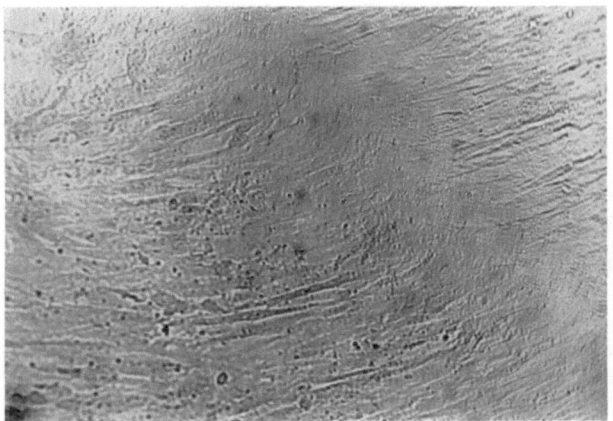

Fig. 1. Uninfected ELF cells

fluid from a 4-day infected culture had a titer of about 10^4 $TCID_{50}$ per ml and, when it was tested in guinea pigs, it contained about 10^6 LD_{50} per ml. Thus, this cell culture was less sensitive to the virus than laboratory animals. However the

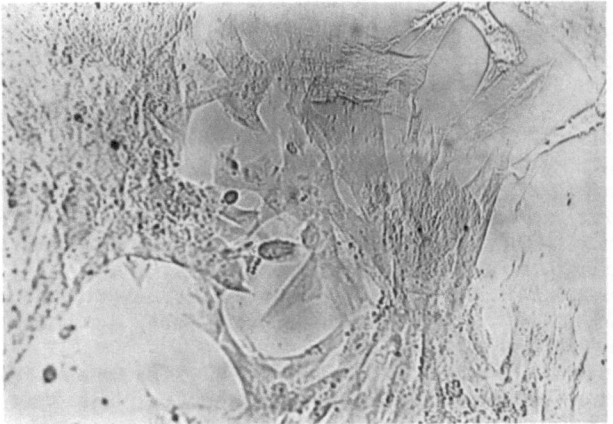

Fig. 2. Beginning CPE

virus gives a clear-cut CPE and endpoints are readily determinable so that this cell line can be useful for further studies of the virus.

Summarizing the sometimes conflicting results obtained by the various investigators, it can be stated that cells deriving from mammals such as monkey, guinea pig, and hamster and human cells are susceptible to the virus. Often high titers were produced by the cells although as a rule no gross cytopathic changes could be observed.

Out of the more than 30 different cell systems tested, it appears that only primary cells (Haas et al., 1968) and a strain of permanent Cercopithecus kidney

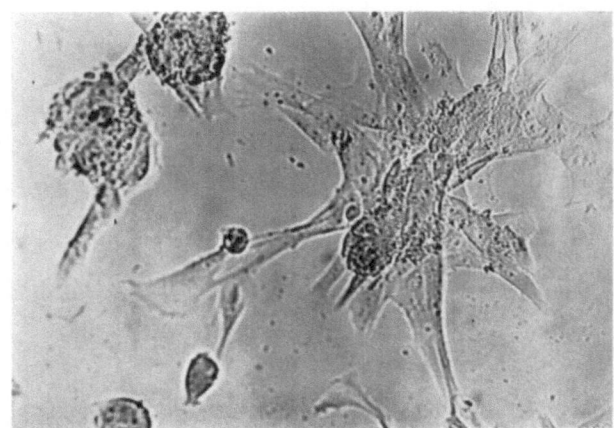

Fig. 3. Maximal CPE

cells (Kissling et al., 1968), BHK$_{21}$ cells (Zlotnik et al., 1968; Kissling et al., 1968) and ELF (embryonal lung fibroblasts) cells exhibit a distinct CPE after infection with the Marburg virus.

References

Haas, R., Maas, G., Müller, J., Oehlert, W.: Experimentelle Infektionen von Cercopithecus aethiops mit dem Erreger des Frankfurt-Marburg-Syndroms (FMS). Z. med. Mikrobiol. u. Immunol. **154**, 210—220 (1968).

Hofmann, H., Kunz, Ch.: Das Verhalten des sogenannten „Marburg-Virus" in einigen Gewebekulturen. Zbl. Bakt. I. Orig. **208**, 344—347 (1968).

Kissling, R. E., Robinson, R. Q., Murphy, F. A., Whitfield, S. G.: Agent of Disease Contracted from Green Monkeys. Science, **160**, 888—890 (1968).

Kunz, Ch., Hofmann, H., Kovac, W., Stockinger, L.: Biologische und morphologische Charakteristika des Virus des in Deutschland aufgetretenen „Hämorrhagischen Fiebers". Wiener Klin. Wschr. **80**, 161—162 (1968).

May, G., Knothe, H.: Bakteriologisch-virologische Untersuchungen über die in Frankfurt/M. aufgetretenen menschlichen Infektionen durch Meerkatzen. Deutsche Med. Wschr. **93**, 620—622 (1968).

May, G., Knothe, H., Hülser, D., Herzberg, K.: Elektronenmikroskopische Befunde bei einer Affenseuche (Cercopithecus aethiops). Zbl. Bakt. I. Orig. **207**, 145—151 (1968).

Siegert, R., Shu, H.-L., Slenczka, W., Peters, D., Müller, G.: Zur Ätiologie einer unbekannten, von Affen ausgegangenen, menschlichen Infektionskrankheit. Deutsche Med. Wschr. **92**, 2341—2343 (1967).

Siegert, R., Shu, H.-L., Slenczka, W.: Nachweis des „Marburg-Virus" beim Patienten. Deutsche Med. Wschr. **93**, 616—619 (1968).

Smith, C. E. G., Simpson, D. I. H., Bowen, E. T. W., Zlotnik, I.: Fatal Human Disease from Vervet Monkeys. Lancet **1967/II**, 1119—1121.

Zlotnik, I., Simpson, D. I. H., Bright, W. F., Bowen, E. T. W., Batter-Hatton D.: Growth of Vervet Monkey Disease Agent in BHK Cell Cultures. Br. J. Exp. Path. **49**, 311—315 (1968).

Passage of Marburg Virus in Guinea Pigs

Y. Robin, P. Brès, and R. Camain

With 5 Figures

Following the outbreak of a mysterious disease among laboratory workers handling Cercopithecus tissue in Frankfurt and Marburg, and at the request of WHO, infectious materials were sent to us to determine whether the causal agent could be yellow fever or another arbovirus.

This is a brief account of our findings.

Samples of our isolate materials have been sent to the Microbiological Research Establishment, Porton, to confirm that they were identical to Marburg virus. We never received the results and we have no definite proof that our isolates are really Marburg virus strains.

Materials and Methods

Materials

Human Samples

Clinical specimens and autopsy material from patients hospitalized at Frankfurt.

Patient HF: Clotted blood 1, taken 2 days after onset of fever (HF 1) and clotted blood 2, taken 6 days after onset of fever (HF 2).

Patient BG: Dead 7 days after onset of fever: liver and spleen.

Patient PS: Dead 4 days after onset of fever: liver, spleen, and brain.

Monkey Samples

From Frankfurt: blood and autopsy material from 8 *Cercopithecus aethiops*. Specimens of these monkeys were pooled according to the degree of liver necrosis pools F 4-7, F 2-3-8, and F 1-5-6.

From Marburg: liver and kidney from 12 *Cercopithecus aethiops* in four pools: M 9, M 10, M 11, and M 14.

Virological Methods

Preparation of Inocula

Clotted blood was haemolyzed with sterile distilled water and the organs triturated using a cold mortar and phosphate buffer diluent containing 10 p. 100 normal rabbit serum added to give a final concentration of about 20 per cent by volume.

Inoculation

Suckling mice were inoculated intracerebrally (ic) and intraperitoneally (ip) with 0.02 ml.

Guinea pigs (200—300 g) were inoculated ip with 4 ml. For further passaging, whole blood harvested by cardiac puncture was injected ip.

Monolayer tube cultures of primary monkey kidney cells (*Erythrocebus patas*) and KB cells were inoculated with 0.1 ml amounts.

Safety Precautions

All operations were carried out in a low pressure hood equipped with ultra-violet lamps and an exhaust system with an air filter. Infected animals were kept in a special room.

Bacteriology

Aerobic and anaerobic media were inoculated to control sterility. Blood cultures of all guinea pigs were negative.

Histological Methods

Table 1 summarizes the different conditions of the disease studied in our guinea pig material.

Necropsy material is fixed in Duboscq-Brasil, embedded in paraffin, and stained by Hematoxylin-Eosin, Mann, Giemsa 55 °C.

Results

All isolations attempts have been negative except in guinea pig.

Guinea pig Passages

Guinea pigs injected by ip route with blood samples of patient H. F. (HF 1 and HF 2) organ suspensions of patient P. S. and Cercopithecus organ pools M 10 and M 14 consistently developed a febrile reaction 4 to 6 days after inoculation. The febrile stage lasted 3 to 7 days.

Whole blood taken during this febrile stage has been successfully passaged ip in guinea pigs through 3 to 6 passages (Figs. 1 to 5). The incubation period was shortened to 2—3 days and some guinea pigs died from 7 to 17 days after inoculation. Clinical symptoms in the animals were: loss of appetite and weight, "bloated face" and enlargement of the testes. At autopsy, we found splenomegaly and lung consolidation.

Fever reactions in guinea pigs were produced using early blood samples from the preceding passage. In one instance (Fig. 4, passage 4), fever reaction followed inoculation of blood taken at the 11th day. This fact confirms observation that infection seems to produce a long lasting viraemia.

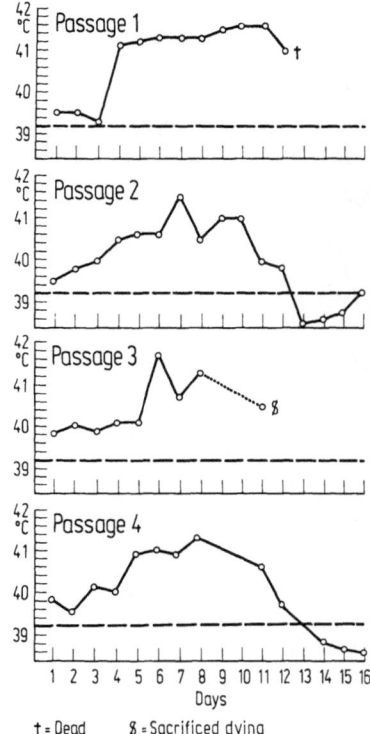

Fig. 1. Blood from patient HF 1

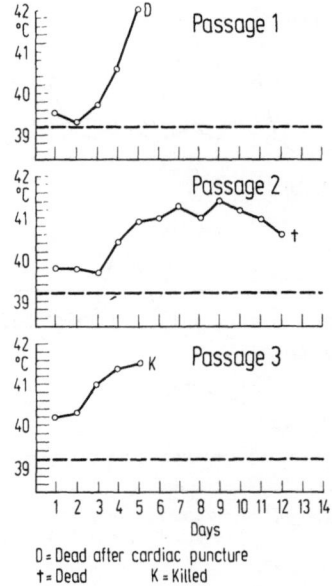

Fig. 2. Blood from patient HF 2

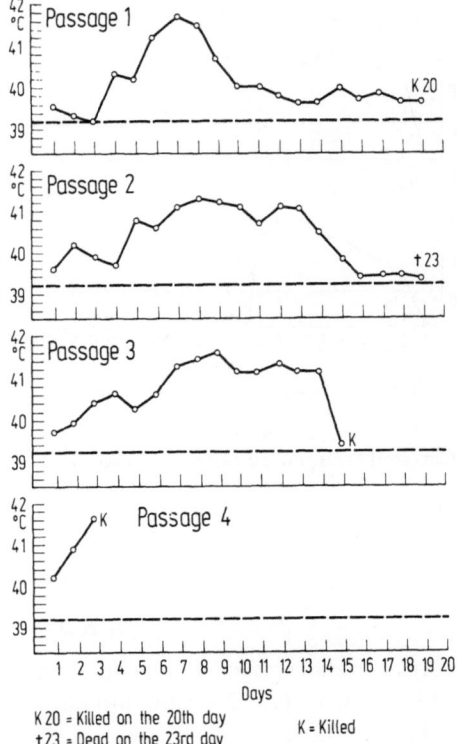

Fig. 3. Post-mortem material from
patient P.S.

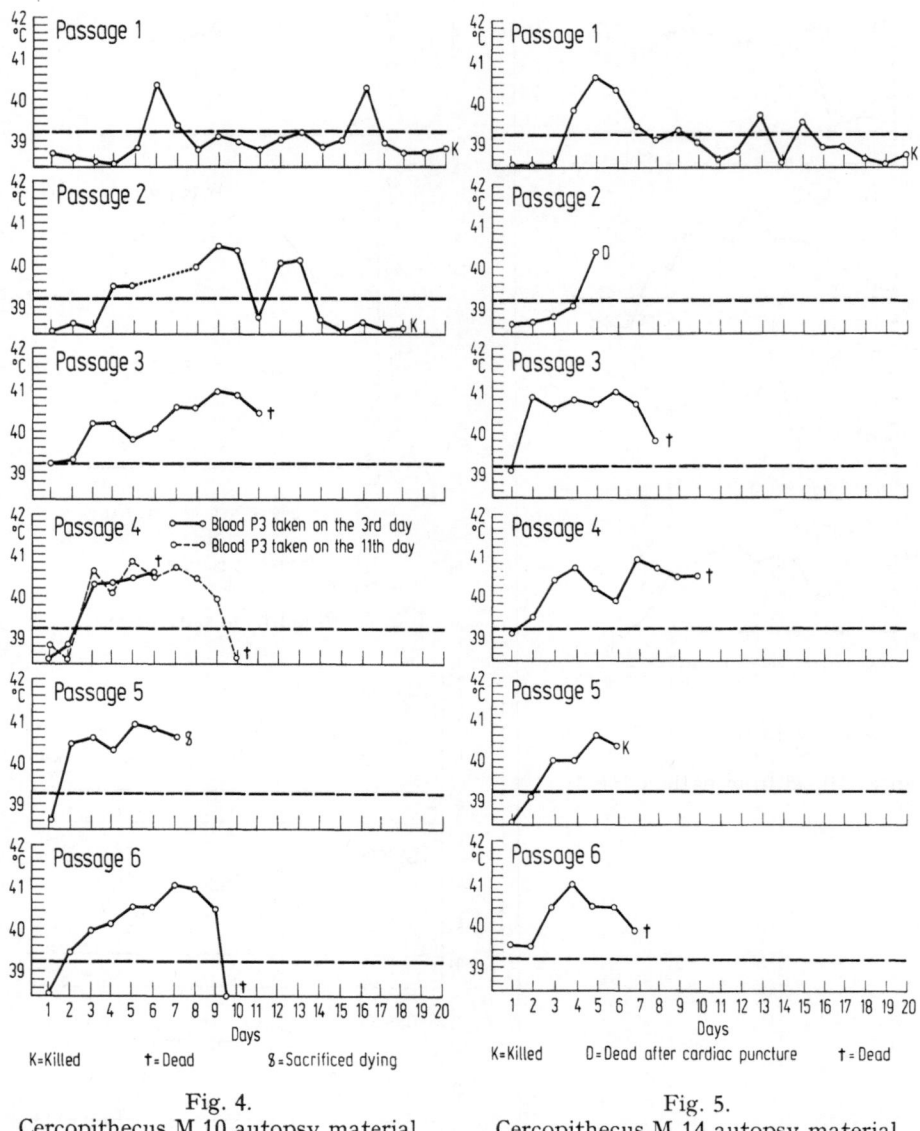

Fig. 4.
Cercopithecus M 10 autopsy material

Fig. 5.
Cercopithecus M 14 autopsy material

Pathology

Microscopic Findings

At a similar period of development these findings were roughly the same.

Liver: On day 3: light interstitial infiltrate; rare liver cells beginning hyaline necrosis.

On days 5-6-7: enhancement of interstitial infiltrate (histiocytes and amphophilic polymorphonuclear leucocytes) surrounding small foci of hepatic cells showing partial or total hyaline necrosis. Some of these degenerate cells look like Councilman bodies of yellow fever.

On days 12 to 15: foci of hyaline necrosis are larger and more numerous, associated with an infiltrate of the same nature as before.

On day 21: small foci of infiltrate, nearly without liver cell necrosis.

Spleen: Days 3—5: congestion of venous sinuses with presence of numerous amphophilic polymorphonuclear leucocytes and sometimes atrophy of lymph follicles (Stress).

Days 6—15: congestion of the red pulp, with presence of amphophilic polymorphonuclear leucocytes, associated with proliferation of different types of R. E. S. cells (free histiocytes, macrophages, marginal cells of venous sinuses).

After staining with Giemsa 55°, organelles of less than 1 μ in size were seen in numerous histiocytes and macrophages (about half or one third the size of one of the eosinophilic granules of amphophilic polymorphonuclear leucocytes). These organelles were seen in four spleens.

Day 21: decrease of amphophilic polymorphonuclear leucocytes but no change in enhancement of R. E. S. cells.

Lungs: The lesions of interstitial pneumonia occur early (day 3) and are seen all along the course of the disease as well as in many cases hypertrophy of juxta bronchic lymph nodules. Occasionally a focus of leucocytic pneumonia is seen.

Other findings:

Kidneys: no special findings.

Heart: moderate interstitial and sub-endocardic infiltrate.

Adrenal glands: were not systematically collected.

In a guinea pig sacrificed when dying (day 12), very severe hyaline necrosis of numerous cortical spongiocytes was observed. This animal presented otherwise very severe liver injuries.

Brain: congestion.

The liver, lung and spleen lesions were seen in all cases. It would be very interesting:

1 To explore the adrenal lesions systematically,

2 To compare the organelles mentioned in spleen histiocytes with the findings of other workers.

Conclusions

The results reported above show that from human and monkey material, an organism has been isolated and transmitted through four to six passages in guinea pigs.

The most striking feature was the presence in spleen histiocytes and macrophages of small bodies less than 1 μ in size.

Table 1.

Passage Days	Hartz I		Hartz II		Popp			Breiter	M14	M10
	1	3	1	3	1	3	4	1	5	5
3							S 41°6			
5			S 42°	S 41°3						
6									S 40°4	
7										S 40°6
12	+									
14		D								
15						AH				
21					AH			AH		

Abbreviations:
+ Dead
S 42° Sick — Temperature 42 °C
D Sacrificed when dying
AH Apparently healthy

Early Histological Lesions — (3rd to 7th day)

Guinea pig inoculated with Popp S. — 4th passage — Sacrificed on 3rd day — T° 41°6.

1 — Liver — Rare foci of lymph cell infiltrate.

2 — Spleen — Congestion — Polymorphonuclear infiltrate.

Guinea pig inoculated with Hartz 2 — 3rd passage — Sacrificed on 5th day — T° 41°3.

3 — Liver — Partial or total hyaline necrosis of some liver cells. When the necrosis is total, the cells look like Councilman bodies.

4 — Spleen — Severe congestion — White pulp nearly empty of lymphocytes (Stress). Polymorphonuclear infiltrate.

5 — Lung — Interstitial pneumonia.

Guinea pig inoculated with M 10 — 5th passage — Sacrificed on 7th day — T° 40°6.

Liver — A focus of liver cell necrosis with interstitial infiltrate.

Spleen — Congestion — Light reaction of RES cells.

Lungs — Interstitial pneumonia.

Heart — Interstitial and subendocardial infiltrate.

Histologic Findings in Livers and Spleens of Guinea Pigs after Infection by the Marburg Virus

G. KORB and W. SLENCZKA

In this paper we report on investigations with the Marburg Virus which were closed shortly before this symposium. Our results originate from 26 guinea pigs, which especially in the 1st, 3rd and 4th, partly also in the 5th, 6th as in the 9th passage were infected intraperitoneally, intravenously or intracerebrally and either died or were killed between the 2nd and 21st day of illness. Our interest was on the one hand to see how far the morphological changes in the livers and the spleens agree with the findings in human beings and on the other hand whether we could discover other peculiarities.

At first it could be seen —as already known—that connections exist between the kind of infection and the climax of the changes. So we found the most severe changes in the early sequence on the 7th and 8th day of illness, in the late passage already on the 5th day of illness.

In the liver we found single-cell necroses which were disseminated and scattered all over the lobules. We constantly demonstrated side by side different stages of the development of the necroses. The single cell necroses were marked by a homogeneous transformation, an increased eosionophilia of the cytoplasm and pyknotic nuclei. Later on they looked like the Councilman bodies. At a greater magnification it was seen that the liver cells showed circumscribed regressive changes of the cytoplasm in the form of condensations or clottings shortly before the total necrosis.

Besides multiple single-cell necroses, group necroses developed. When they only covered a few liver cells, an activation of the Kupffer cells always appeared very early; when there was a larger area affected, it first came to a sudden decay of the necrotic liver cells. After this a considerable proliferation of the Kupffer cells filled the raised defects. Inflammatory cell collections could seldom be proved within the sphere or in the vicinity of the necroses. Here and there liver-cell mitoses existed in the vicinity of the necrotic areas. In all cases with liver necroses, a fatty degeneration of the liver cells was to be seen at the same time. The periportal tracts showed only a slight inflammatory reaction with monocytoid elements. Cholestasis could be seen in no case. In the livers of the animals killed on the 17th or on the 21st day of illness, the only sign of liver damage was a circumscribed proliferation of Kupffer cells. Therefore we conclude that these changes may be reversible. As the changes in the guinea pig livers were chiefly marked by disseminated single-cell and group necroses of liver cells with a conspicuously slight inflammatory reaction, an early proliferation of Kupffer cells and a more or less marked fatty degeneration, there is a remarkable correspondence with the findings in the *human* liver caused by this infection (BECHTELSHEIMER 1968, GEDIGK, BECHTELSHEIMER and KORB 1968, KORB, BECHTELSHEIMER and GEDIGK 1968).

Besides this, in five cases we could find small basophilic corpuscles—similar to the human material—partly in the periportal tracts, occasionally in the liver cells. Rarely small eosinophilic bodies surrounded by narrow uncoloured areas could be seen in hepatocytes. At present it is not possible to explain their importance.

Other liver cells contained granular or irregular basophilic structures which showed positive PAS and v. Kossa reactions. Just the same structures were described by SIMPSON, ZLOTNIK and RUTTER (1968) and ZLOTNIK (1969). They consider that there are connections between the basophilic structures and accumulations of the infective agent.

Finally, we found here and there so-called fibrinoid thrombi inside the sinusoides and enlarged Kupffer cells which had phagocyted a hyaline fibrinoide material.

Even in the spleen, the changes resembled the human material. Thus we could prove multiple partial follicular necroses; at the same time we found a slight but significant amount of cells in the red pulp on the 7th day of illnes. The red pulp contained a moderate eosinophilic, partly homogeneous, partly more granular material. In a further phase we found at first in circumscribed areas and later on diffuse multinuclear giant cells as well as reticular and monocytoid cells in the red pulp. Corresponding changes were found in one animal on the 17th day of illness and a beginning fibrosis of the pulp could be seen in another animal on the 21st day of illness.

In 10 of the 26 guinea pigs we saw basophilic corpuscles in the spleens, chiefly between the 6th and the 9th day of illness—similar to those we described in the liver. They were partly diffusely scattered in the red pulp. Finally, we could identify cells which contained a hyaline fibroid material.

Summary: it follows from our investigations that morphological changes in the liver and the spleen of guinea pigs caused by an infection of the Marburg Virus show a remarkable conformity with the findings in human beings. Further peculiarities were discovered, but the explanation can only be found by new experiments and by the use of additional methods.

References

BECHTELSHEIMER, H.: Die pathologische Anatomie der „Marburg-Virus"-Krankheit. Habil. Schrift, Med. Fak. Univ. Mbg., 1968.

GEDIGK, P., BECHTELSHEIMER, H., KORB, G.: Die pathologische Anatomie der „Marburg-Virus"-Krankheit (sog. „Marburger Affenkrankheit"). Dtsch. Med. Wschr. **93,** 590—601 (1968).

KORB, G., BECHTELSHEIMER, H., GEDIGK, P.: Die wichtigsten histologischen Befunde bei der „Marburg-Virus"-Krankheit. Ärzt. Mitt. **65,** 1089—1096, (1968).

SIMPSON, D. J. H., ZLOTNIK, J., RUTTER, D. A.: Vervet monkey disease. Brit. J. Exper. Path. **49,** 458—464 (1968).

ZLOTNIK, J.: Marburg agent disease: Pathology. Roy. Soc. Trop. Med. Hyg. **63,** 310—323 (1969).

Neuropathological Findings in Experimentally Infected Guinea Pigs

H. Solcher

With 5 Figures

We examined neuropathologically 15 guinea pigs infected with the Marburg-Virus.

The animals were inoculated intraperitoneally and intracerebrally and became ill 7 to 12 days after injection. The duration of illness was 2—9 days (Table 1). There were no serious neurological findings. After preparation with formalin and

Table 1.

Nr.:	Duration of illness	Encephalitis	Immunofluorescence
	intraperitoneal inoculation		
129	8 days		+
131	8 days		+
132	7 days	+	+
133	8 days		+
139	7 days		+
67	9 days	+	not examined
68	8 days		not examined
81	8 days	+	not examined
82	7 days		not examined
	intracerebral inoculation		
95	2 days		∅
96	2/4 days	+	∅
97	3 days		∅
114	5 days	+	(+)
115	6 days		∅
116	7 days		∅

embedding in paraffin, the brains were cut in series. About every 10th slice was colored with Cresylviolet or Haemalaun-Eosin. In negative cases we colored in closer sequence.

Five out of 15 brains showed glial-nodule encephalitis; 2 of the animals were inoculated intracerebrally, 3 intraperitoneally. The glial nodules are sparse, in 2 cases even rare. They are found in the grey and white matter, but definitely more often in the grey matter.

A detailed topical pattern cannot be set up because of the small number of glial nodules. The process was, however, mainly located in the medulla and mid-brain (Fig. 1). The lack of changes in the cerebellum in all 5 cases is remarkable.

In the beginning of the process there is only an assembly of some microglial cells ("Stäbchenzellen") (Fig. 2).   This glial reaction can be seen without any vascular-inflammatory reaction. In the further progress of the disease this primary glial reaction is followed by diffuse glial cell proliferation or spotted glial nodules

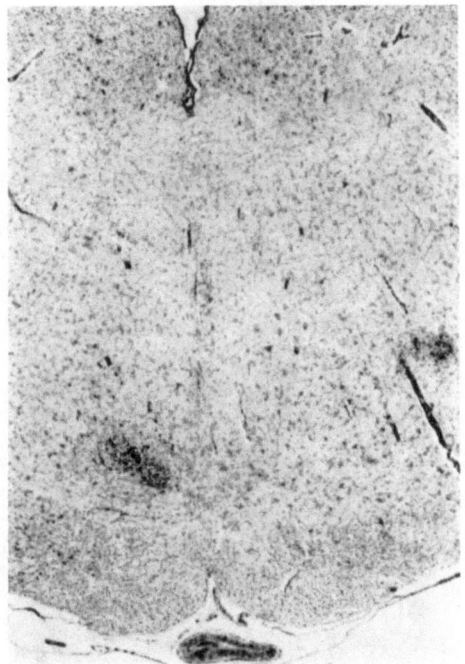

Fig. 1.   Glial nodules in the brainstem

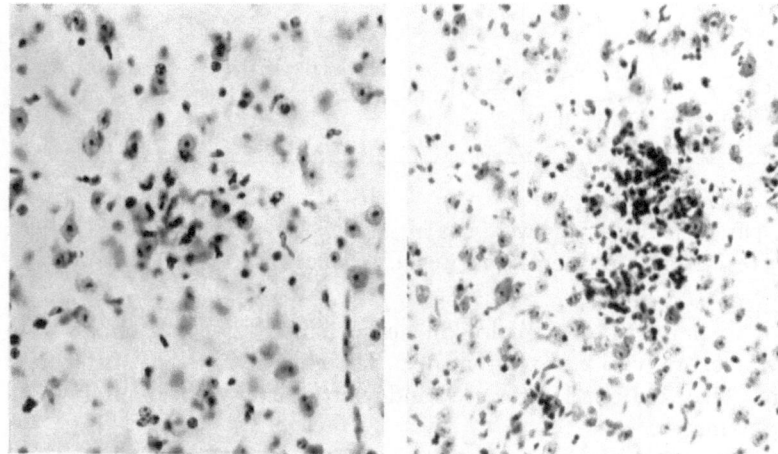

Fig. 2. Assembly of microglial cells   Fig. 3.   Spotted glial nodule
            in the cortex   in the cortex

(Fig. 3). Here microglia is also dominating over macroglia. In the center of larger nodules we have often seen regressive changes (Fig. 4). There are no irritative and regressive changes on the neuron and no neuronophagies.

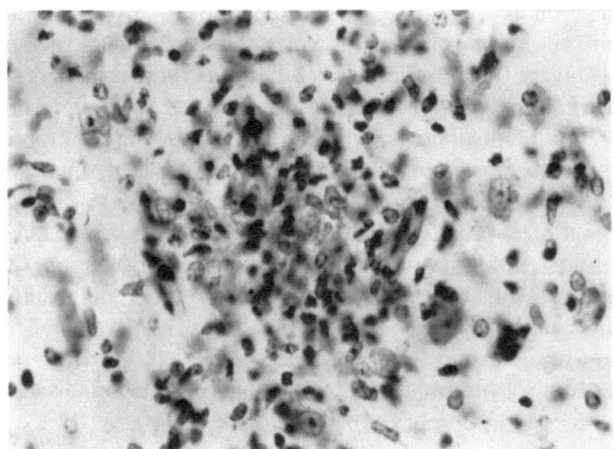

Fig. 4. Regressive changes in the center of a nodule

The mesenchymal reaction is sparse. Only near some glial nodules perivascular infiltrations are seen (Fig. 5). Nevertheless endothelial hyperplasia may sometimes simulate a larger inflammatory reaction. Also meningeal inflammatory

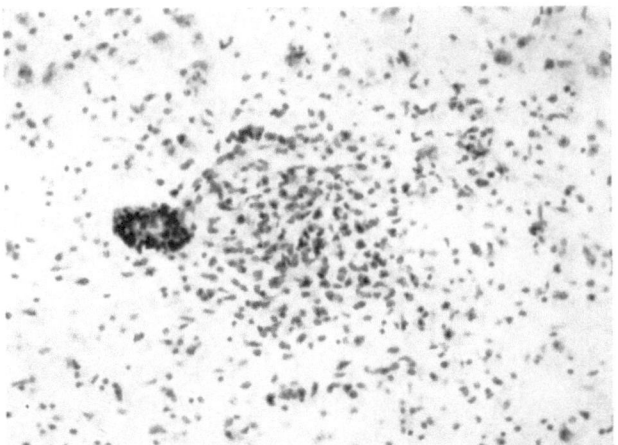

Fig. 5. Perivascular infiltration near a glial nodule

reactions show a lymphocytic irritation. Occasionally around those inflammatory changed vessels there may be intracerebral and meningeal haemorrhages.

In guinea pigs the Marburg Virus infection shows the findings of a glial-nodule encephalitis of polioencephalitic type with preference of tihe brainstem.

The topic resembles the Lyssa, Epidemic and Borna Encephalitis. In Lyssa and Epidemic Encephalitis there are serious inflammatory findings in the beginning, while the Borna Encephalitis begins with microglial proliferation before the appearance of perivascular round-cell-infiltrates. Seifried and Spatz (1930) pointed to this especially.

Compared with the human cases investigated by us in 1968 (Jacob and Solcher) there is a topical shift of the glial process. In humans there is always a panencephalitis with equal findings in the white and grey matter without preference of the brainstem or cerebral hemispheres, while in guinea pigs there is a polioencephalitis with preference of the brainstem. This topical shift we see as a species-specific variation as is known from research on encephalitis.

It seems to us of special interest that there is no correlation between our histological findings and the immunofluorescence findings of Dr. Slenczka (Hygiene-Institute of Marburg University) who found a spotted fluorescence in the brain of guinea pigs (Fig. 1).

We also found no neuropathological changes in 3 baby mice inoculated intracerebrally in the first 24 hours of life, which survived several weeks and showed immunofluorescence in the cerebrum.

References

Jacob, H., Solcher, H.: Über eine durch grüne Meerkatzen übertragene, zu Gliaknötchenencephalitis führende Infektionskrankheit („Marburger Krankheit"). Acta neuropath. **11,** 29 (1968).

Seifried, O., Spatz, H.: Die Ausbreitung der encephalitischen Reaktion bei der Bornaschen Krankheit der Pferde und deren Beziehung zu der Encephalitis epidemica, der Heine-Medinschen-Krankheit und der Lyssa des Menschen. Z. Neur. **124,** 317 (1930).

"Marburg Disease"
The Pathology of Experimentally Infected Hamsters

I. Zlotnik

With 10 Figures

Introduction

Very soon after its first appearance in Germany as a human disease entity, the agent of Marburg disease has been transmitted to guinea-pigs (Smith et al. 1967; Seigert et al. 1967; May and Knothe, 1968) and to monkeys (Simpson et al., 1968 a & b and Haas et al., 1968). Recently the range of susceptible animals has been increased by the successful transmission of the agent to hamsters (Simpson, 1969). To begin with only suckling hamsters proved to be susceptible and the disease was evident in only 40—80 per cent of animals after 5 to 11 days following either i.c. or i.p. inoculations. Successive passaging caused a reduction in the incubation period and a mortality of almost 90 per cent of infected hamsters. After nine passages in sucklings, 5—6 weeks old hamsters were also inoculated and the majority developed signs of illness and mortality 5—8 days after i.c. injections and 6—10 days after i.p. inoculations. The pathological changes in organs of infected animals resembled those of guinea pigs and monkeys, but CNS lesions hitherto unobserved in other animals were constantly present in i.c. inoculated hamsters (Zlotnik, 1969 and Zlotnik & Simpson, 1969).

Macroscopic Changes

Macroscopic lesions were not readily noticeable in organs of hamsters inoculated with guinea-pig or monkey material. After five successive passages in suckling hamsters however, striking changes were observed in the liver and brain, while less marked naked eye changes were seen in the spleen, kidney, and lung. Adult hamsters inoculated with 9th suckling hamster passage material had lesions similar to those of sucklings, but the CNS lesions were confined only to adult hamsters inoculated i.c.

The liver was usually enlarged, congested, soft, and either dark purple or brick red in colour. The brain of all i.c. inoculated animals and i.p. injected sucklings was always soft, reddish, or pink and studded with a variety of haemorrhages. The spleen was only rarely enlarged, but was usually hard and purplish in colour. The kidneys were congested, while the lungs appeared either normal or (about 50 per cent) had dark red foci of consolidation.

Histological Changes

1. Liver

As a rule liver lesions became progressively more obvious and marked as the number of passages increased. Most livers of hamsters after the 5th passage had widespread lesions consisting of degeneration and necrosis of hepatocytes with the

formation of cellular debris, swelling of Kupffer cells and the appearance of large
free macrophages laden with nuclear fragments. Many liver cells were undergoing a
hyaline degeneration resulting in Councilman-like bodies. The degeneratiye process was accompanied by an attempt at regeneration and liver cells showing mitotic
figures were frequently seen. Apart from the degenerative changes in liver cells
mononuclear cells resembling reticulum cells infiltrated the organ and formed
large accumulations in the periportal spaces and occasionally in the vicinity of
central veins (Fig. 1).

Beginning from the 9th hamster passage large clumps of basophilic granules
appeared within the cytoplasm of some liver cells. The individual bodies were
$\frac{1}{2}$—$1\frac{1}{2}$ μ in diameter, were round, elliptical or ring shaped and appeared to be
surrounded by a matrix. The bodies proved to be weakly, Feulgen positive,
stained reddish-purple with Giemsa and bright red with Machiavello. They were
brown when stained with von Kossa, fluorescent in U.V. light, strongly PAS positive, and did not stain with Gomori's aldehyde fuchsin. Incubation at 60 °C in
normal HCl for 10 minutes abolished von Kossa's reaction, but did not affect the
PAS reaction (Fig. 2).

2. Spleen and Lymph Nodes

Intensive and widespread proliferation of reticulo-endothelial tissue was a
constant feature in all spleens and in some lymph nodes. This was usually accompanied by the formation of large numbers of free macrophages which tended to
accumulate in the sinuses. After eight passages in hamsters in addition to proliferative changes in the spleen, degeneration and necrosis of lymphoid cells within
the Malpighian bodies became marked (Fig. 3).

3. Kidneys

Limited parenchymatous degeneration of the tubular epithelium and congestion were invariably present in the cortex, but shrinkage of the glomerular tufts
and thickening of Bowman's capsule were only occasionally seen. In addition, in
hamsters of the 9th and 10th passage, the renal medulla contained foci of mononuclear cellular accumulations and clumps of various sizes composed of basophilic
granules situated either in the renal epithelium or in the capillary endothelium.
Very often the clumps of intracytoplasmic bodies were surrounded by large
infiltrations of mononuclear cells. The granules exhibited similar histochemical
properties to those present in the liver (Figs. 4 & 5).

4. Lung

The affected lungs showed evidence of focal interstitial pneumonitis with
cuboidal metaplasia and proliferation of vascular endothelium. The alveolar sacs
contained varying numbers of polymorphs and macrophages laden with cellular
debris. The capillaries were very often distended and contained coagulated blood,
but occasionally also fibrinous clots. In a few instances, in lungs of hamsters of 9th
and 10th passages accumulations of intracytoplasmic granules were seen either
in the alveolar epithelium or in the capillary endothelium. These granueles caused
as a rule great distention of the cellular cytoplasm (Fig. 6).

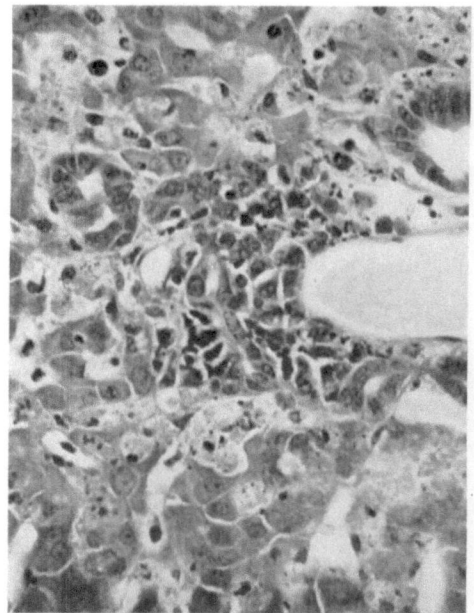

Fig. 1. Liver—Necrosis and periportal accumulation of mononuclear cells. H & E., ×350

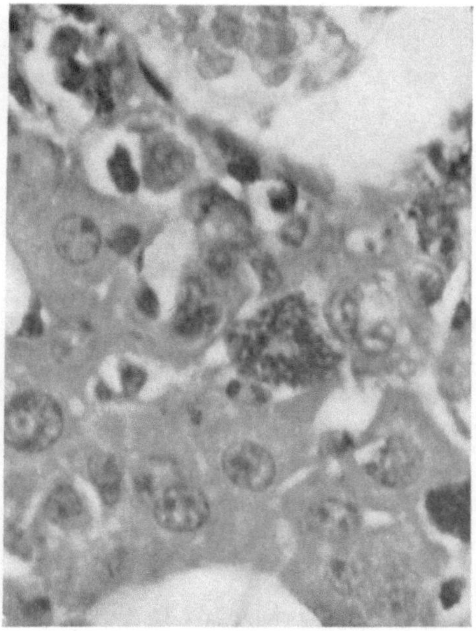

Fig. 2. Liver—Intracytoplasmic bodies clumped together within a liver cell. H & E., ×900

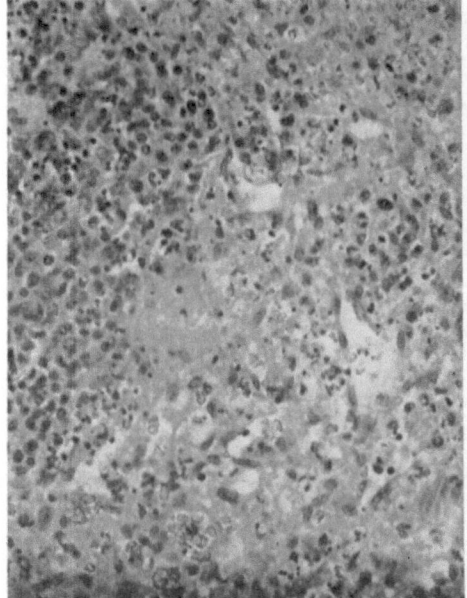

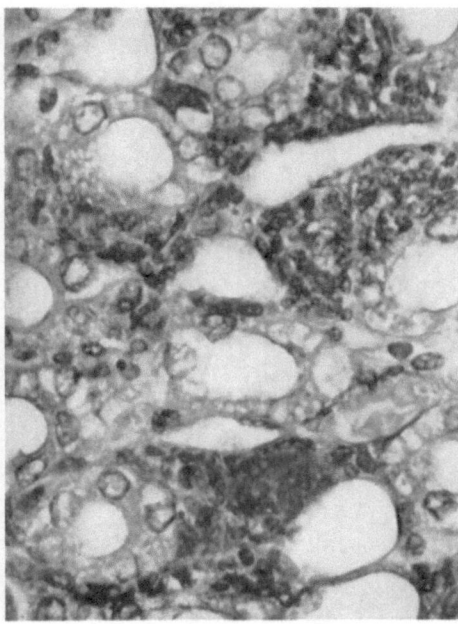

Fig. 3. Spleen—Necrosis within a Malpighian body, H & E., ×350

Fig. 4. Kidney—Large infiltrations of mononuclear cells within the medulla. H & E., ×630

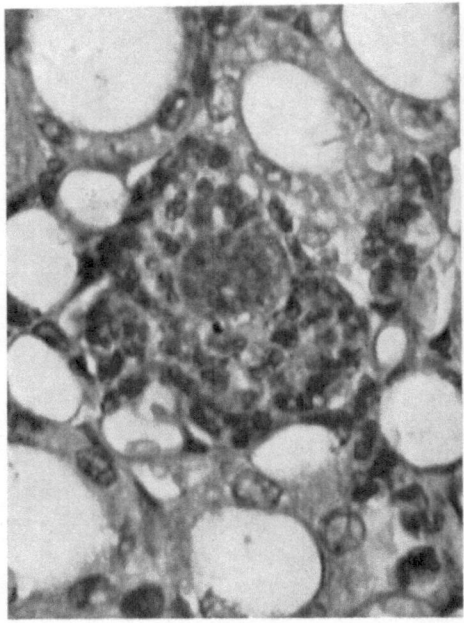

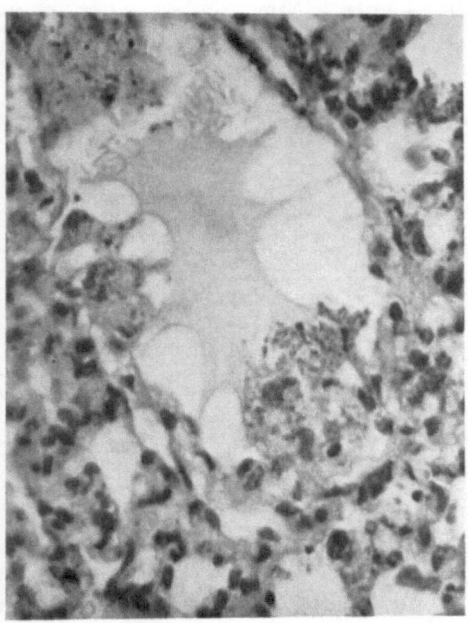

Fig. 5. Kidney—Intracytoplasmic bodies clumped together and surrounded by mononuclear cells within the renal medulla. H & E., ×900

Fig. 6. Lung—Large accumulation of intracytoplasmic bodies within the vascular endothelium. H & E., ×630

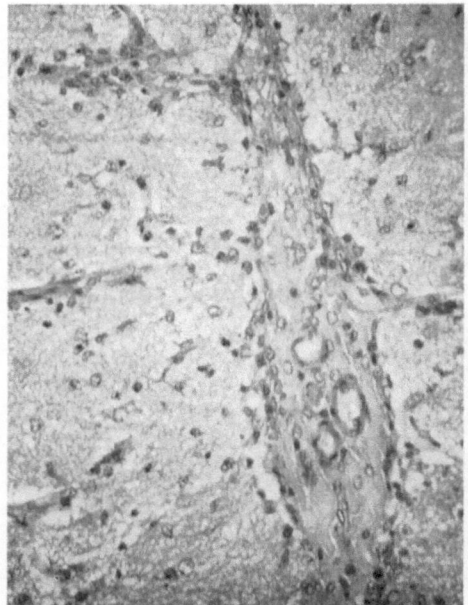

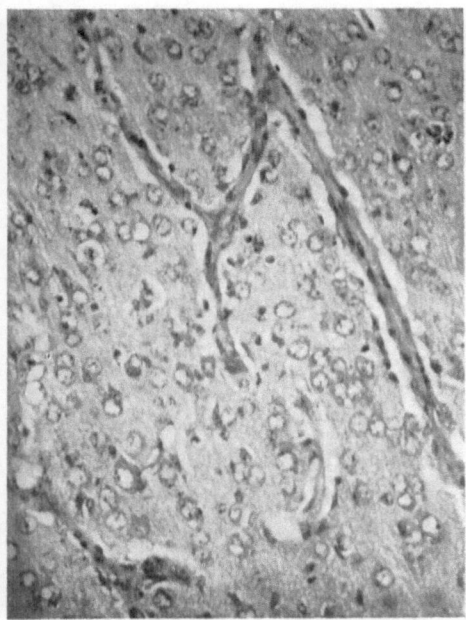

Fig. 7. Brain—Meningeal exudate and perivascular infiltrations. H & E., ×220

Fig. 8. Brain—Diffuse microglial proliferation and new capillary formation. H & E., ×220

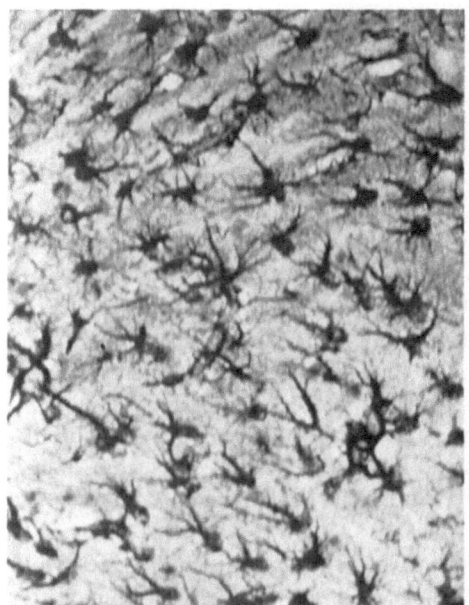

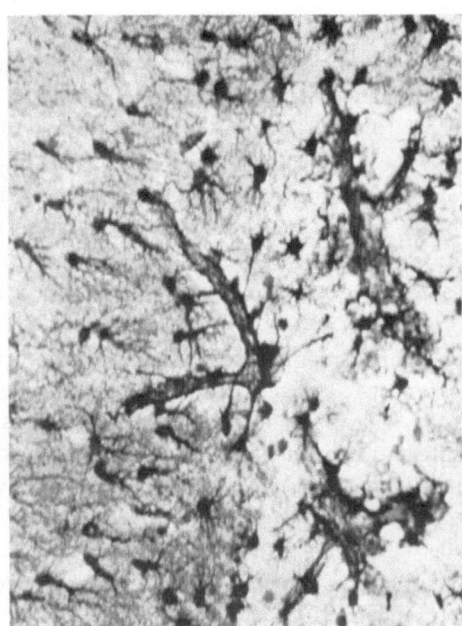

Fig. 9. Brain—Astrogliosis in the hippocampus. Cajal., ×325

Fig. 10. Brain—Peri-capillary haemorrhages and astrocytic proliferation. Cajal., ×280

5. Brain

Suckling and adult hamster inoculated intracerebrally usually developed lesions of meningoencephalitis. The infiltrating cellular elements consisted of microglial and mononuclear cells that resembled the infiltrating cellular elements in the liver and kidney, but at no time were lymphocytic perivascular cuffings observed. The capillaries were proliferating and the endothelium showed increased cellularity. Astrocytic hypertrophy and multiplication was very widespread and marked especially around the capillaries in the hippocampus and at the periphery of the cerebral cortex. Perivascular haemorrhages were extremely common throughout the brain and varied from small pericapillary accumulations of red corpuscles to very large extravasations of blood giving rise to malacia, necrosis and neuronal degeneration. In addition to the above lesions, of special interest was severe necrosis of all brain elements around the needle track after intracerebral inoculations. In many instances large areas of the cortex, hippocampus, and thalamus were undergoing complete necrosis with very severe astrogliosis at the periphery of the necrotic zones (Figs. 7—10).

Brains of hamsters inoculated by the intraperitoneal route had lesions similar to animals infected intracerebrally except that they had no necrotic lesions connected with the needle track. On the other hand in adult hamsters inoculated intraperitoneally, the brain changes observed consisted only of small pericapillary haemorrhages.

Discussion

The transmission of the agent of Marburg disease to hamsters is of special significance in that it became adapted to the central nervous system and at the same time increased the range of susceptible laboratory animals.

The pathological changes in the hamster are similar in some respects to those observed in other animals and in man, but the tendency to produce widespread brain lesions in suckling hamsters is so far unique as it is not confined to those inoculated only by the intracerebral route but also by intraperitoneal injection BECHTELSHEIMER et al. 1968, GEDIGK et al., 1968, HAAS et al. 1968, SIMPSON et al. 1968, SMITH et al. 1967, ZLOTNIK, 1969, and ZLOTNIK and SIMPSON, 1969).

In the liver of infected hamsters, apart from widespread degeneration, necrosis, attempts at regeneration of hepatocytes, and focal infiltrations of mononuclear cells, clumps of intracytoplasmic bodies similar to those present in the liver of infected guinea-pigs, appeared in hamsters of 9th and 10th passage. Similar structures were present also in the kidneys and lungs and in this respect the hamster is different from the guinea-pig, where intracytoplasmic bodies were found only in the liver.

The intracytoplasmic bodies seen under the light microscope and referred to previously as V. M. D. bodies (ZLOTNIK, 1969) formed clusters of basophilic granules or ring-shaped structures $\frac{1}{2}-1\frac{1}{2}$ μ in diameter, which were cemented together by a matrix that showed histochemical properties common to gluco- or mucoproteins and had a tendency for calcium salt deposition. It is possible that these intracytoplasmic bodies represent the causative agent of the disease (SMITH et al. 1967). The absence of such bodies from the livers of man, monkey, first passage guinea pigs, and approximately eight hamster passages and their appearance in guinea pigs from the second passage onwards and hamsters from 9th passage, may be due to a chemical change in the matrix surrounding them during the course of passaging which rendered them visible in the light microscope.

The proliferation of reticulo-endothelial cells and the depression and degeneration of lymphoid elements was remarkable and noticeable not only in the spleen and lymph nodes, but also in the central nervous system where the inflammatory changes were completely devoid of perivascular lymphocytic cuffings. The haemorrhagic changes in the brain merit special attention as they seem to be the primary lesion and a spring-board for vascular changes, astrocytosis, and degeneration.

Summary

Lesions caused by Marburg disease in hamsters, although resembling in many respects those observed in man, monkey, and guinea-pig are different in that the intracytoplasmic bodies believed to be the causative agent of the disease appeared after nine passages, not only in the liver as in the guinea pig, but also in the kidney and lung.

Encephalitis is a regular feature in all suckling hamsters, irrespective of route of inoculation and in adult hamsters after i. c. injection. The process in the brain consists of microglial and astroglial proliferations, haemorrhages, and neuronal degeneration, but at no time perivascular lymphocytic cuffings are present.

References

BECHTELSHEIMER, H., JACOB, H., SOLCHER, H.: Dt. med. Wschr. **93,** 602 (1968).

GEDIGK, P., BECHTELSHEIMER, H., KORB, G.: Ibid., **93,** 590 (1968).

HAAS, R., MAASS, G., OEHLERT, W.: Medsche. Klin. **35,** 1359 (1968).

HAAS, R., MAASS, G., MULLER, J., OEHLERT, W.: Z. med. Mikrobiol. Immunol. **154** 210 (1968).

MAY, G., KNOTHE, H.: Dt. med. Wschr. **93,** 620 (1968).

SIEGERT, R., SHU, H.-L., SLENCZKA, W., PETERS, D., MULLER, G.: Dt. med. Wschr. **92,** 2341 (1967).

SIMPSON, D. I. H., ZLOTNIK, I., RUTTER, D. A.: Br. J. exp. Path. **49,** 458 (1968a).

SIMPSON, D. I. H., BOWEN, E. T. W., BRIGHT, W. F.: Lab. Anim. **2,** 75 (1968b).

SIMPSON, D. I. H.: Br. J. exp. Path. **50,** 389 (1969).

SMITH, C. E. G., SIMPSON, D. I. H., BOWEN, E. T. W., ZLOTNIK, I.: Lancet **1967ii,** 1119.

ZLOTNIK, I.: Trans. R. Soc. trop. Med. Hyg., **63,** 310 (1969).

ZLOTNIK, I., SIMPSON, D. I. H.: Br. J. exp. Path. **50,** 393 (1969).

Experimental Infection of Monkeys with the Marburg Virus[1]

R. Haas and G. Maass

With 9 Figures

We are indebted to Dr. Siegert who kindly provided our starting material of three blood specimens. Two of these were taken from human subjects during the acute phase of the disease. A guinea pig provided the other specimen, after 4 passages of the causative agent in this species.

Our first experiment was to establish: 1) whether the Marburg virus is pathogenic for monkeys, 2) whether (and, if so, how) it is transmitted among monkeys, and 3) the disease pattern and pathological changes it brings about in monkeys. Rhesus and vervet monkeys were inoculated. Some of the latter had passed through London on the way to our institute around the same time as had those implicated in the outbreaks of disease in Marburg and Frankfurt.

Fig. 1 shows the housing of the monkeys. Each monkeys was in a separate cage. We removed the isolating partitions separating pairs of cages to provide

Fig. 1. View of cages for the maintenance of monkeys during the experiments

the possibility for contact between pairs of monkeys through the lattice work. One monkey only, of each such pair, was infected experimentally, the other not being treated at all. In the same room we also placed four Rhesus monkeys,

[1] The investigations reported here were carried out in collaboration with Drs. Maass and Oehlert.

1 —2 meters from the infected monkeys, which were not inoculated. These served as controls to establish whether the causative agent could be transmitted over a certain distance, as, for example, by droplet infection.

The results of the first experiment are shown in Table 1. The extreme right hand column shows that all monkeys—Rhesus or vervet—which were injected with blood died between the 7th and 9th day after injection. Further, all control

Table 1. *Infection of Cercopithecus aethiops and Macaca rhesus with "Vervet agent"*

Group	Inoculated material	Inoculated monkeys	Death (days p.i.)
A	Human blood Cl*	2 *M.rhesus*	9, 9
	Controls	2 *M.rhesus*	18, 18
	Human blood Cl*	2 *C.aethiops*	8, 9
	Controls	2 *C.aethiops*	19, 26
B	Human blood HO**	2 *C.aethiops*	9, 9
	Controls	2 *C.aethiops*	18, 20
C	Guinea pig blood***	2 *M.rhesus*	7, 8
	Controls	2 *M.rhesus*	16, 16
	Guinea pig blood***	2 *C.aethiops*	7, 7
	Controls	2 *C.aethiops*	15, 36
D	Room contacts	4 *M.rhesus*	76 surviving

* 1.0 ml s. c.
** 0.5 ml s. c.
*** 1.0 ml 4th guinea pig passage.

animals, which had direct contact with the injected monkeys, died between the 15th and 36th day. On the other hand, the four rhesus monkeys living in the same room but without direct contact with the infected monkeys, showed no symptoms of disease after 76 days, when the experiment was ended. On the average, the rhesus controls died after 17 days, the vervet controls surviving 20 days (or 22 days, if one which died after 36 days is included).

Dr. OEHLERT will report in the pathological findings revealed at autopsy.

Blood specimens were taken immediately before and one and six days after infection. Examination of these specimens is still being carried on.

The following conclusions can be drawn from the first experiment:

1. The causative agent of the Frankfurt—Marburg syndrome is pathogenic for rhesus as well as vervet monkeys.

2. This agent is transmitted where direct contact between infected and susceptible animals obtains. There is no evidence that spread of this agent is dependent upon a vector, such as an arthropod. Apparently the infection is not possible without direct contact.

Superficial inspection of the animals did not reveal any impressive clinical symptoms: one or two days before death the monkeys appeared weak and sickly. Usually they sat huddled apathetically in their cages, did not eat, and reacted only slightly to their environment. Skin eruptions were not seen. Death occurred relatively quickly without noticeable symptoms. Fig. 2 shows average temperature curves, based on daily measurement from the second day on, of the infected and control groups, but they do not reveal any valuable information.

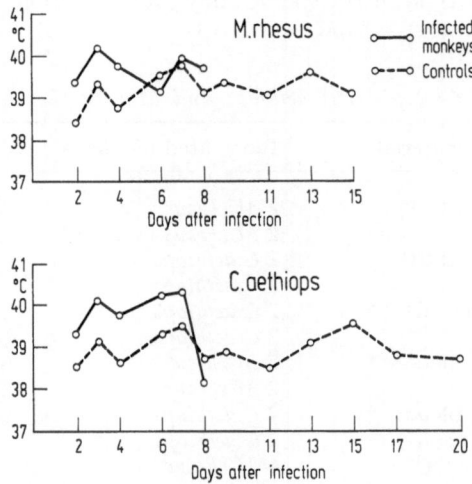

Fig. 2. Medium temperature of infected and non-infected monkeys

In the first reports which reached us concerning the disease in Marburg and Frankfurt, it was evident that the human infections occurred predominantly following contact with the blood of vervet monkeys. Because of this, we attempted

Table 2. *Titration of a blood sample of 1st Cercopithecus passage of blood Ho**

Dilution**	Death (days p.i.)
10^0	5, 7
10^{-2}	7, 7, 6
10^{-4}	7, 9
10^{-6}	8, 10, 10
10^{-8}	11, 12
10^{-10}	14, 25

* The blood sample was taken at the 10th day after infection of a Cercopithecus with blood Ho.
** 1.0 ml s. c.

to determine the infectiousness of the blood of a vervet monkey for other vervet monkeys, shortly before its death. To this end blood was withdrawn from a monkey on the 10th day after it had been inoculated with infectious human (Ho)

blood—the day before its death, as it turned out—and maintained at minus 60 °C
for a few weeks until testing.   Table 2 shows the results of the titration experiment
for infectiousness of the blood.   One ml of each dilution was injected subcutaneously into monkeys.   Up to the dilution of 10^{-10} all animals died, although one
survived 25 days before it finally died.   At autopsy all showed the changes typical
of Marburg virus infection.   A frank prolongation of the time interval between
injection and death was observed only at dilutions of 10^{-6} and greater.   The experiment shows that the blood of a monkey infected with Marburg virus can
contain enormous quantities of the causative agent 24 hours before its death.
In passing it is interesting to note that a titration of the same blood specimen
which had such a high degree of infectiousness for vervets, showed on primary
vervet kidney cell culture a titer of approximately $10^{8.0}$ $TCID_{50}/ml$.

The specific etiology of this disease was not established with the observation
that the injection of acute phase blood from a Frankfurt-Marburg Syndrome
patient elicits a fatal disease with certain histopathological changes in monkeys.
Besides the histological examinations Dr. OEHLERT, the etiology of this disease as
manifested in monkeys was investigated in 4 other ways:

a) By direct electron microscopic demonstration of the agent in the blood of
infected monkeys (investigations undertaken by Dr. PETERS and his colleagues);

b) By electron microscopic demonstration of virus particles in the livers of
infected monkeys;

c) By specific demonstration of virus antigen in liver, spleen, and lymph nodes
of infected monkeys, using the fluorescent antibody technique;

d) By monolayer tissue culture of cells from kidneys taken from infected
monkeys followed by demonstration of Marburg virus infection using cytological
or fluorescent antibody techniques.

Fig. 3 shows a section of liver from a Marburg virus-infected vervet monkey.
Specimens of liver tissue were fixed with osmium tetroxide, embedded in
Vesopal and contrasted with uranyl acetate.   These investigations were carried

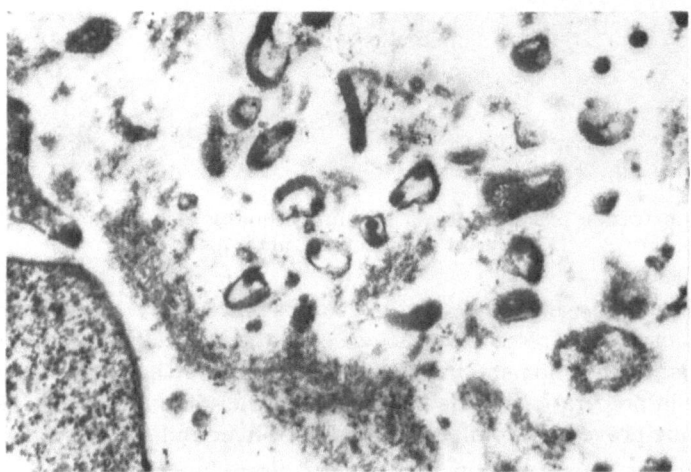

Fig. 3. Marburg virus in an ultrathin section of liver cells from an infected monkey

out by Dr. Müller in our laboratory. We found virus particles in the livers of each of the three monkeys examined. They were observed to be mostly in the cytoplasm, but also extracellular. Frequently we saw structures resembling triangles, where several virus particles appeared to lie in a capsule. Single cells contained, besides the characteristic virus particles, large, more or less homogeneous masses with a diameter of 2—3 microns. Occasionally tube- or cord-shaped structures were seen. Possibly these are the electron microscopic equivalent of eosinophilic inclusion bodies, as seen with the light microscope, which we did not observe in the livers of non-infected vervets.

We could demonstrate virus antigen specifically with the fluorescent technique in sections of liver, spleen, and lymph nodes, as well as in swab preparations from liver and spleen. Dr. Mauler of Marburg kindly provided 8 human convalescent sera, which we pooled for our antiserum. Acetone-fixed sections 5 μ in thickness were treated with this pooled antiserum after diluting 1 : 10. For the negative control serum we used a pool of 8 randomly chosen sera from our virology diagnostic section. Dr. Vogt of our laboratory prepared the rabbit anti-human gamma globulin serum coupled with FITC, which was used to demonstrate the specific antibody localization.

Fig. 4 shows the specific greenish fluorescence in a liver section. This fluorescence, which we were regularly able to demonstrate in the organs of infected

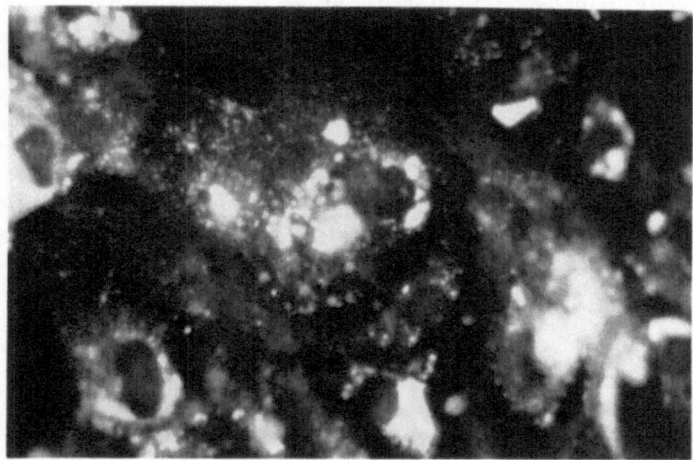

Fig. 4. Demonstration of viral antigen by immunfluorescence technique in section of liver of an infected C.aethiops

monkeys, always appeared to be spot- or band-shaped in the cytoplasm. We did not observe nuclear fluorescence. Also, accompanying specific fluorescence, there was always abundant non-specific fluorescence. Fig. 5 shows the virus-specific fluorescence in preparations made from swabs of the spleen of an infected vervet. This technique proved uniformly successful with liver and spleen swabs.

Since—to our knowledge—in Marburg and Frankfurt there were persons who contracted the Frankfurt Marburg Syndrome solely by working with primary tissue cultures derived from vervet kidneys, we decided to make further experiments with such cultures. Specifically, we tried find out whether the Marburg virus infection could be demonstrated cytologically or with the fluorescent technique.

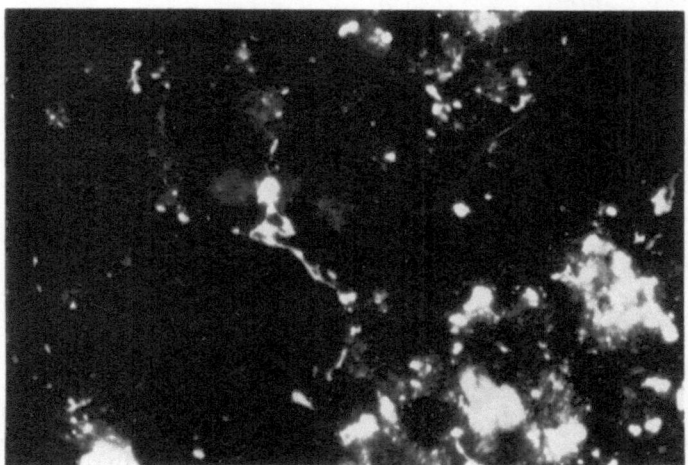

Fig. 5. Demonstration of viral antigen by immunfluorescence technique in a smear of the spleen of an infected C.aethiops

Accordingly, 6 vervets were infected experimentally by subcutaneous injections of 1 ml. of blood of a vervet monkey, which had been previously infected with Ho blood. This blood was taken from the monkey one day before its death. For preparing monolayer kidney cell cultures from the 6 monkeys, we used the trypsin digest procedure of Bodian, in order to avoid unnecessary risks of infection. With this method the pieces of kidney tissue are treated for 12 hours at 4 °C with trypsin, without changing the enzyme solution and without washing the cells free of enzyme. It was our plan to kill the animals at the first sign of symptoms, to remove the kidneys and to prepare cell cultures from them. Two animals died before we could carry out the plan. As expected, the kidney cell cultures derived from these two casualties showed unsatisfactory growth and did not form a confluent cell layer. However, these, as well as the other four monkeys which were killed and nephrectomized, showed the histopathological legions in liver, spleen and lymph glands typical for the Marburg virus infection.

The tissue cultures derived from the kidneys of these four monkeys showed, in comparison to those derived from kidneys of non-infected monkeys, relatively normal growth. A cell monolayer which was practically confluent developed in a comparable period of time. With the usual low-power microscopic method of observation, no unequivocal cytopathological effect could be established within 2—3 weeks. In a few cases these cultures could even be subcultured once.

However, if one stains the primary cultures and examines them at higher magnification, cytoplasmic inclusions similar to those seen in the livers of infected monkeys, can be seen in many cells (Fig. 6). We were able to observe them by the second day after planting. The number of cells with these inclusions varied among the four monkeys. In cultures from two of the animals we could see them in nearly all cells, while in those cultures from the other two, only a few cells possessed them.

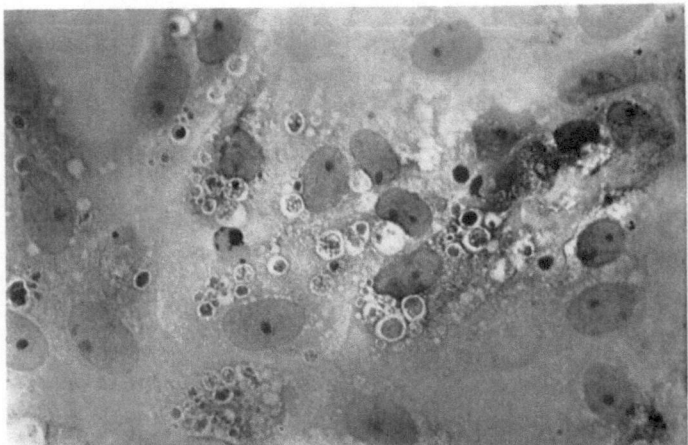

Fig. 6. Kidney tissue cultures from an infected C.aethiops killed during the incubation period of the illness

We could also demonstrate the presence of the Marburg virus in the tissue cultures derived from the kidneys of the four infected monkeys in thin sections with electron microscopy (Fig. 7). Finally, using the fluorescent antibody tech-

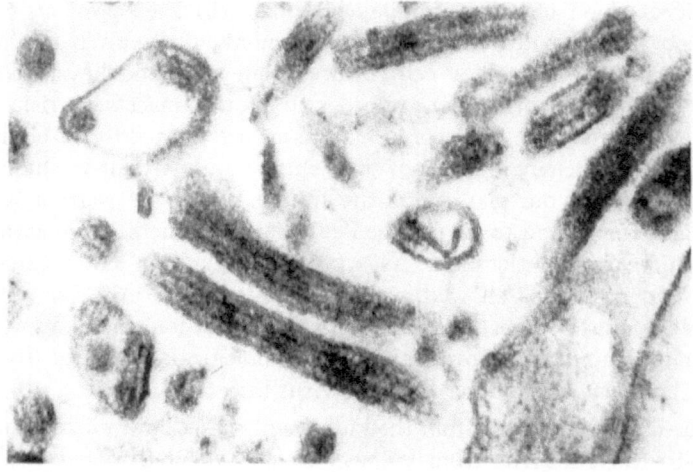

Fig. 7. Marburg virus particles in ultrathin section of an infected tissue culture of monkey kidney cells

nique, Marburg virus antigen was easily demonstrable in primary cultures, as well as in subcultures, of these kidney cells (Fig. 8). Fig. 9 shows normal human serum for comparison.

In conclusion I would like to point out that the Marburg virus can be cultivated in stationary monolayer cultures of vervet monkey kidneys. The infection can be

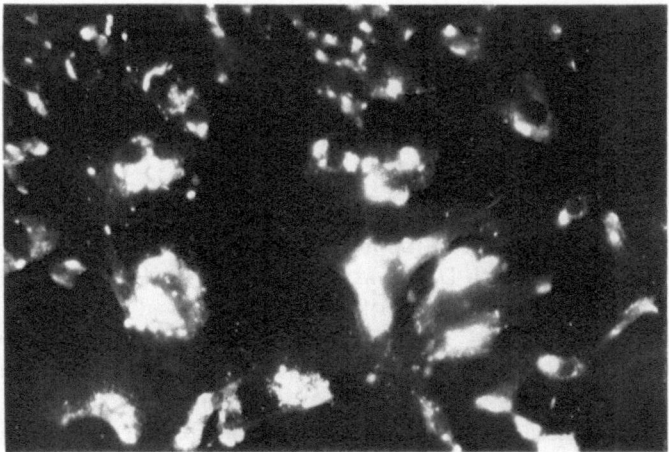

Fig. 8. Demonstration of Marburg virus antigen by immunfluorescence in infected kidney cell culture

recognized by the cytopathological effect which, in our experience never involved the entire cell lawn.

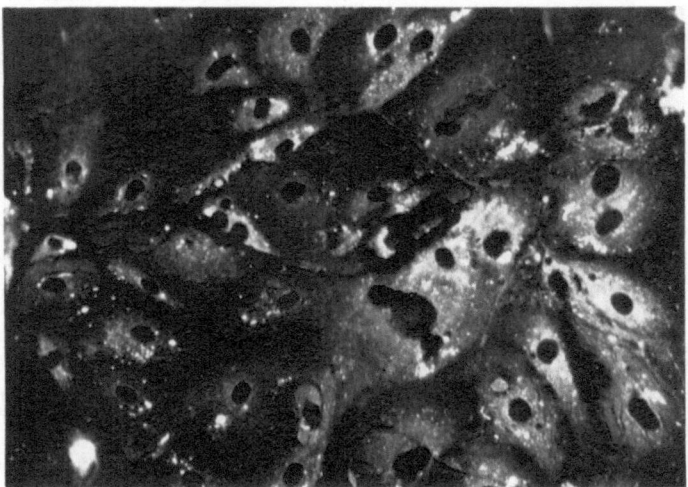

Fig. 9. Negative control

Our investigations have shown that the Marburg virus is highly pathogenic for rhesus and vervet monkeys, that it is directly transmissible from one monkey to another by intimate contact, and that its presence can be established in different ways, including the culture of tissue derived from infected monkeys.

The Morphological Picture in Livers, Spleens, and Lymph Nodes of Monkeys and Guinea Pigs after Infection with the "Vervet Agent"

W. OEHLERT

With 14 Figures

The experiments with rhesus monkeys and vervets (*Cercopithecus aethiops*) performed in cooperation with Dr. MAASS and Prof. HAAS [1, 2, 3] (Hygiene Institute, Freiburg) were designed to clarify the chronological course and nature of organ and tissue alterations. Eight monkeys were infected with human and guinea-pig blood samples of the fourth passage. All directly infected animals died within 7 to 9 days after inoculation. The cage neighbours of infected animals developed the same disease, however after longer incubation time. These animals died within 18 to 36 days (see Table 1). The clinical symptomatology of all infected

Table 1. *Experimental infection of M. rhesus and C. aethiops with human serum and blood of guinea pigs after passages with "Vervet agens"*

Group	Infectious Material		Animal	Death (days after inf.)	
A	Human serum (1 ml)	2	*M. rhesus*	9,	9
	Cage contact	2	*M. rhesus*	18,	18
	Human serum	2	*C. aethiops*	8,	9
	Cage contact	2	*C. aethiops*	19,	26
B	Human serum (0,5)	2	*C. aethiops*	9,	9
	Cage contact	2	*C. aethiops*	18,	20
C	Blood of guinea pigs	2	*M. rhesus*	7,	8
	in the 4th passage	2	*C. aethiops*	7,	7
	Cage contact	2	*M. rhesus*	16,	16
	Cage contact	2	*C. aethiops*	15,	36

animals was quite uncharacteristic. One or two days prior to death they developed fatigue, refused food, and sat slumped motionless in their cages. Without any additional symptoms, death occurred suddenly. Already on the second day after subcutaneous infection the rectal temperature rose to 40 °C. Autopsy revealed extensive bleeding in the muscles, subpleurally, subendocardially, epicardially and in the mucous membranes of the bladder. The rhesus monkeys showed widespread caseations in livers, spleens, and lungs which macroscopically already could be identified as tuberculosis with beginning cavitation. In the vervets, on the other hand, hemorrhagic suppurative pneumonia was the most striking finding. In both monkey species the liver was almost regularly enlarged, soft, and friable. Similarly the spleen and the para-aortic, parapancreatic, and inguinal lymph nodes were altered.

After considering all histologic alterations which ensued after infection of monkeys with human and guinea-pig blood, we could differentiate between unspecific and cellular and tissue damage specific for this serum infection.

Caseous or productive tuberculosis in the lungs, livers, and spleens of rhesus monkeys, and malaria, worm and bacterial infections of vervet have to be regarded as unspecific alterations. As a matter of fact, the possibility has to be taken in account that preexisting infections could have been activated and influenced in their histologic pattern by this specific viral infection. In contrast to these unspecific organ and tissue alterations there was typical cellular and histologic damage, particularly in the parenchyma of the liver and in the reticuloendothelial system with an almost identical histologic pattern.

We will first describe the hepatic alteration. In addition to the animals listed in Table 1, a number of monkeys had been given subcutaneous inoculation and liver biopsies had been performed after different periods of time. This enabled us to record early cellular alterations, too. Two days after inoculation we observed regularly structured liver tissue with completely intact plate of liver cells. However, rigorous evaluation of the slides revealed some necrobiotic Kupffer cells and occasionally mitoses in intact Kupffer cells. The parenchyma appeared normal. Three days after infection, round, intracytoplasmatic, eosinophilic inclusions which

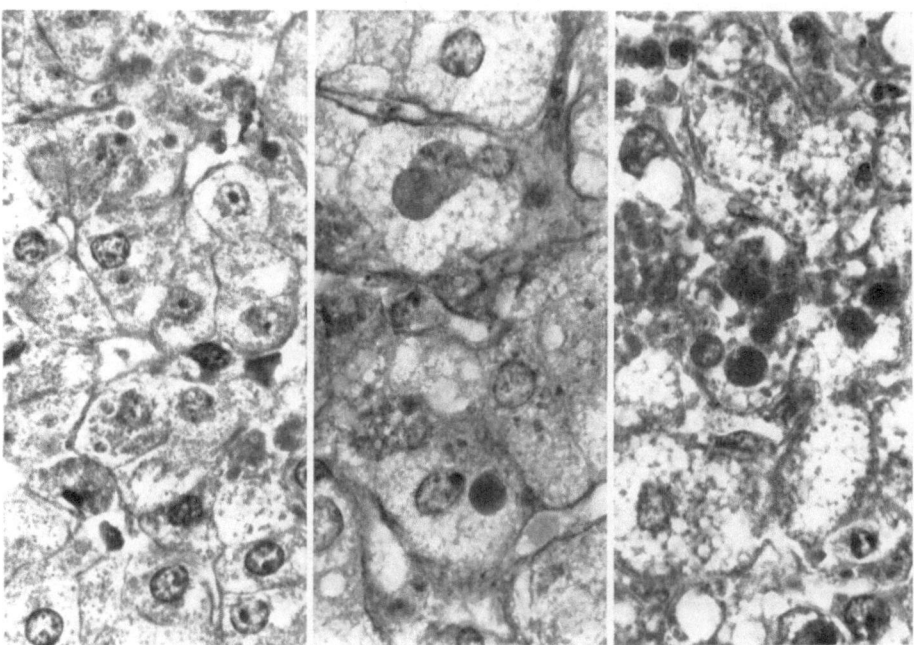

Fig. 1.   Liver of Cercopithecus aethiops 3 days after infection with human serum. Intracytoplasmatic inclusion bodies like focal necrosis

were stained somewhat reddish by PAS stain developed, generally located first in the center of a lobule (Fig. 1). Four days after infection, in addition to these intracytoplasmatic alterations, disseminated cell necrosis could be detected and numerous globular-shaped liver cells were ejected into the enlarged sinusoids (Fig. 2). Following this process, a development of group necroses takes place, frequently surrounded by hemorrhage (Fig. 3). These group necroses present as

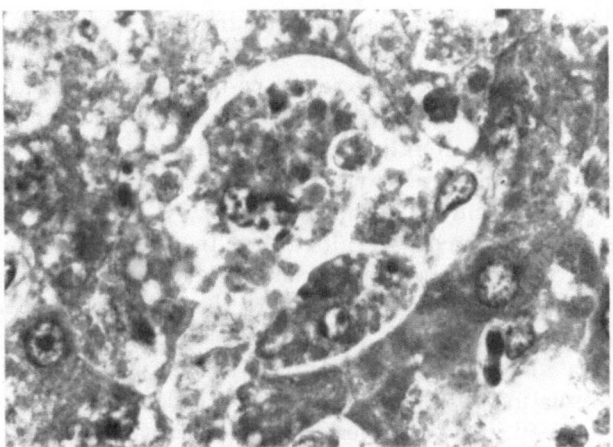

Fig. 2. Liver cell necroses 4 days after Infection of Cercopithecus aethiops with human serum

structures very similar to Councilman bodies in human virus hepatitis. In the neighbourhood of such destroyed cells in the liver parenchyma, hyperplastic Kupffer cells are encountered phagocyting decaying nuclei and globular-shaped

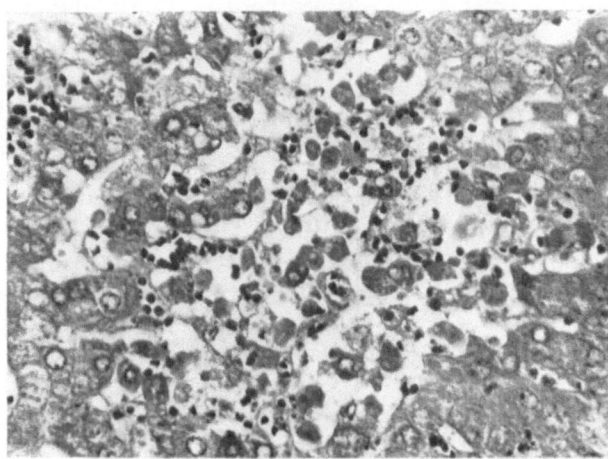

Fig. 3. A group of necrotic liver cells with surrounding redblood cells without inflammation or reaction of RES

eosinophilic material. Only rarely mitoses of parenchyma cells or hyperplastic Kupffer cells are seen. In addition to the partial decay of parenchyma in many animals, spotty or diffuse steatosis of the liver takes place. Therefore some necrotic liver cells or Councilman bodies must still contain lipoids. It has to be stressed that even after very late biopsies, i.e. in the group of subcutaneously infected monkeys after 7 or 8 days and in the group of contact-infected monkeys after 17 to 35 days, no essential proliferative activity was evident in the liver parenchyma and in the reticuloendothelial system.

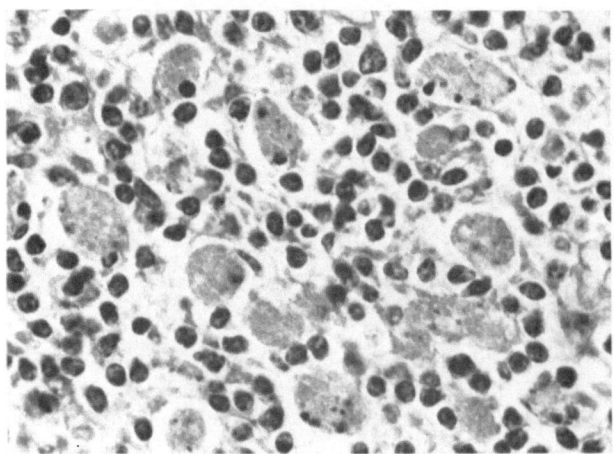

Fig. 4. Parapancreatic lymphnode of Cercopithecus aethiops. Death 9 days after infection with human serum. Swollen reticulum cells with necrobiotic alterations

Even more impressive than those in the liver were the histologic changes in the spleen and lymph nodes investigated by us. In these organs activation of RES was evident with a distinct increase of sinus endothelial and reticulum cells. In later experiments a remarkable packing was seen in the sinus of spleen and lymph nodes of hyperplastic and swollen reticulum cells displaying all phases of decay (Fig. 4), the spleen sinuses, in particular, being filled with detritus of necrotic reticulum cells mixed with a loose fibrinous network. While at first central necroses of the follicles developed (Fig. 5), practically no follicles could be detected in later

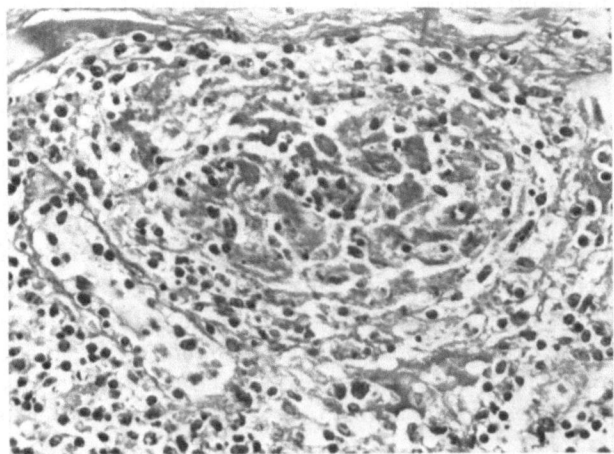

Fig. 5. Spleen of Cercopithecus aethiops. Death 9 days after infection with human serum. Necrosis of the germ centre in a follicle

experimental periods and the only sustained structures in spleen tissue were represented by the trabeculae and sinuses (Fig. 6). As in the liver, no cellular proliferation took place in the spleen and lymph nodes.

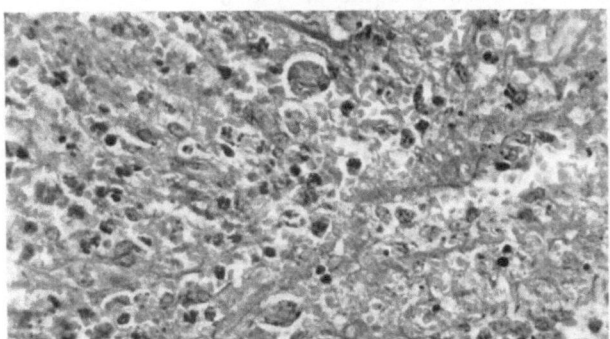

Fig. 6. Same animal. Necrotic material in the red pulp of the spleen

Only rarely we could demonstrate similar intracytoplasmic inclusions and focal necroses as in the liver. No tissue alterations were seen in the pancreas. The lungs of the animals revealed extensive, mostly hemorrhagic focal pneumonia which certainly was one of the main factors causing the death of the animals. In some cases fibrinous casts in the vasa afferentia and capillaries of the kidneys could be demonstrated.

According to these histologic findings, we can comprehensively establish that: a) by means of subcutaneous inoculation or contact infection, this agent can produce focal and group necroses of the liver parenchyma and cause extensive destruction of the RES, especially in livers, lymph nodes, and spleens in *Macacus rhesus* and *Cercopithecus aethiops*. At first we thought that the focal intracytoplasmic inclusions in the form of peculiar eosinophilic globules appearing shortly after inoculation presented a typical histologic pattern for this type of infection. b) We expressed the presumption that a relation may exist between these intracytoplasmatic inclusions and those viral conglomerates in liver epithelial cells demonstrated by other investigators [5]. This presumption is supported by the results of fluorescent microscopic investigations of antibody incubated cells [3, 6].

To clarify the problems, we inoculated 20 guinea pigs in a further survey by intramuscular injection with contagious serum of diseased monkeys. Another group was inoculated with serum from those guinea pigs and in the fourth passage the first symptoms of disease, such as temperature and shagginess of coat, appeared. These animals were chosen for further study. In addition to the usual histological checks of different organs, an autoradiographic investigation was performed after injection of tritiated cytidin, a precursor of RNS and DNS. In addition, tissue material was embedded in Epon and semi-thin and ultra-thin slices were made. The diseased animals were decapitated 3, 4, 5, 6, and 7 days after intramuscular injection.

Already on the third day after inoculation all animals developed temperatures up to 39.8 °C.

Histological Lesions in Livers, Spleens, and Lymph Nodes of Guinea Pigs

In the liver of the experimental animals, as in the infected monkeys, already three days after infection focal and group necroses of parenchymal cells could be demonstrated, presenting homogeneous anuclear structures which looked like the Councilman bodies seen in human degenerative liver diseases (Fig. 7). In contrast

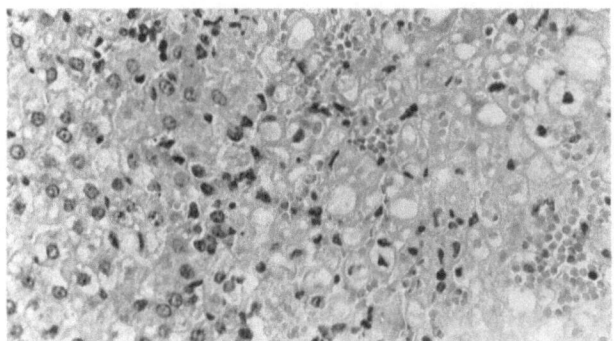

Fig. 7. Liver of guinea pig 3 days after infection with the "Vervet agens" of the 4th passage in guinea pigs. Multiple necroses of liver cells with surrounded bleedings

to the monkeys, however, already at this time distinct focal activation of RES with increase of Kupffer cells ensued. The Kupffer cells with enlarged nucleus very often formed loose nodules.

Four days after infection, the necrotic areas in the liver with anucleous structures and structures looking like Councilman bodies had expanded, some necrotic parenchyma cells had been ejected into enlarged sinusoids and had been phagocytized by hyperplastic Kupffer cells. Furthermore, peculiarly swollen liver epithelials with focal cytoplasmic homogenization were encountered. No such

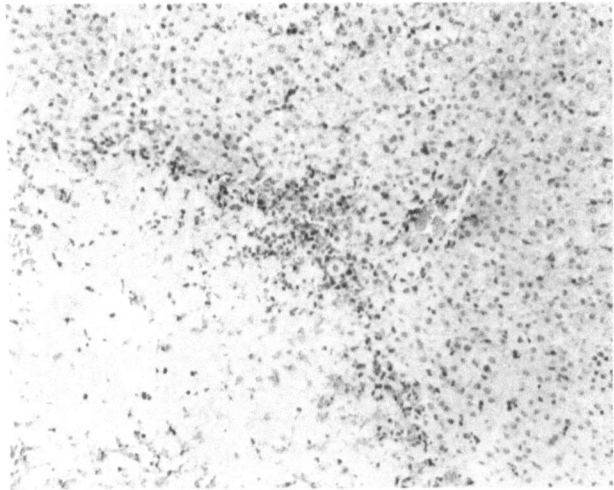

Fig. 8. Liver of guinea pig 5 days after infection. In the periphery of the large necrose proliferation of mesenchymal cells

circumscribed cytoplasmic homogenizations as in the monkeys could be demonstrated in this group. Occasionally polymorphnuclear leucocytes could be found surrounding the foci of necroses, though reaction of RES with general proliferation and formation of loose nodules of Kupffer cells predominated (Fig. 8). At this time no mitoses can be demonstrated in the liver parenchyma. Five days after

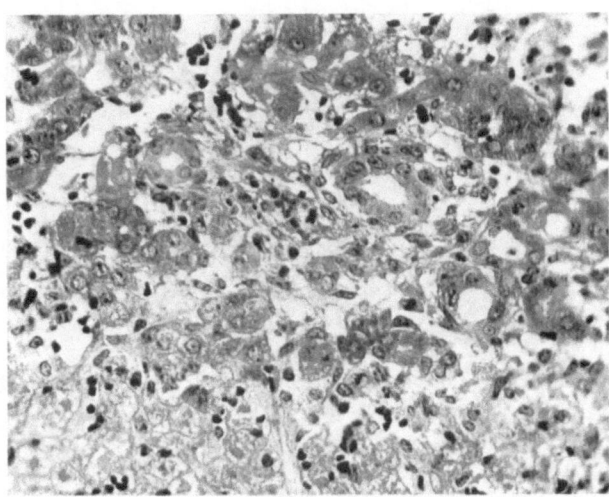

Fig. 9. Liver cell regeneration with so called bile duct proliferations in the periphery of a large necrose zone in liver of guinea pig 7 days after infection

infection necroses have expanded further and one sees map-like areas with nuclear debris and a peripheral mesenchymal reaction separating foci of necrosis and unaltered liver parenchyma.

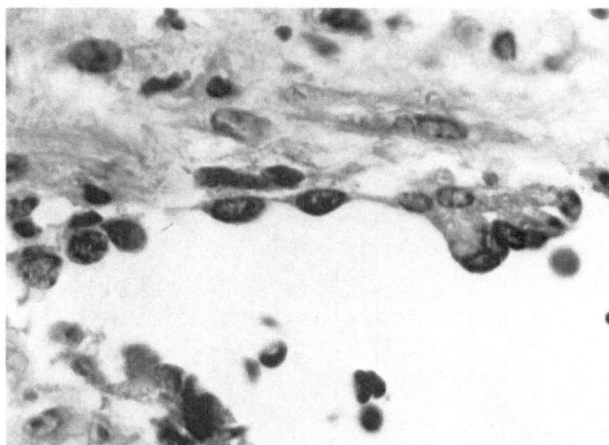

Fig. 10. Proliferation of endothelial cells with multinuclear giant cells in portal vein of a liver of guinea pig 7 days after infection

Six days after infection, regeneration of liver parenchyma takes place heralded by an increase of mitoses; bile duct proliferation is present and solid formations of regenerated liver cells can be observed.

Seven days after infection the necrotic areas are separated from the intact parenchyma by broad strands of collagen-producing cells with spindle-shaped nuclei. Bile duct proliferations are abundant, penetrating the necrotic areas (Fig. 9). A striking finding at this time is extensive endothelial proliferation in hepatic blood vessels with formation of polynuclear endothelial giant cells (Fig. 10).

To obtain further information on the course of cell damage within the first periods of the experiment, 0,25 μ slices were made and stained with a combined PAS stain. Normally in the thin slice the glycogen-filled liver cell presents flattened edges to the neighbouring cells. In between the formation of parenchyma cells, liver sinusoids with their stretched endothelial cells can be visualized.

Intracytoplasmic homogeneous map-like areas are visible in the liver cells stained slightly reddish by our stain, indicating glycogen deposits.

Blue-stained inhomogeneous intracytoplasmic RNA-enriched areas surround the nucleus and are located at the inner side of the nuclear membrane, both localizations being connected by small bridges. The nucleus is mostly stained dark blue, whereas the chromatin is not clearly demonstrable by this stain. 3 and 4 days after infection, the liver cells of our experimental animals revealed in the glycogen-enriched parenchyma in numerous thin slices a light, rounded swollen cytoplasm where complete loss of intracellular glycogen had taken place, accompanied by the probable destruction of the cellular structures localized there. The nucleus of these cells appears swollen and displays a very loose structure of the chromatin. The chromatin is closely attached to the nuclear membrane and the vacuolized nucleolus is surrounded by a small shell of dark-stained nuclear chromatin. Some altered epithelial cells are already ejected from the liver cell formations and can be observed, rounded and swollen, in the lumina of the enlarged sinusoids (Fig. 11a, b, c).

In the thin slices an impressive activation of the RES is present. Virtually all Kupffer cells are hyperplastic and reveal a rounding of formerly spindle-shaped nuclei. The lumina of the liver sinus are filled with those cells. However, it cannot be decided whether these cells derive from Kupffer cells, endothelial cells or the vessels of mononuclear macrophages. Identical activation is present in the vessels of the periportal spaces. The lumina of these vessels are surrounded by hyperplastic endothelial cells which present different stages of decay.

By autoradiography after injection of tritiated cytidin, the activation of RES is documented morphologically by increased uptake of tritiated cytidin in its cells. Under normal conditions the autoradiographical number of the silver grains, as a measure of the activity of cytidin uptake in RNA, is nearly equal. In our experimental animals, however, a higher density of silver grains is encountered over the nuclei of Kupffer cells. The previously mentioned light liver epithelials cells with swollen nucleus reveal a significantly decreased uptake of cytidin compared with their neighbouring cells. Frequently no activity at all can be demonstrated in

these cells with light cytoplasm, so that one can assume that already at the time
of the experiment the synthesis of RNA had stopped in these cells (Fig. 12).
In samples taken later, the map-like areas of necroses appeared as light spaces

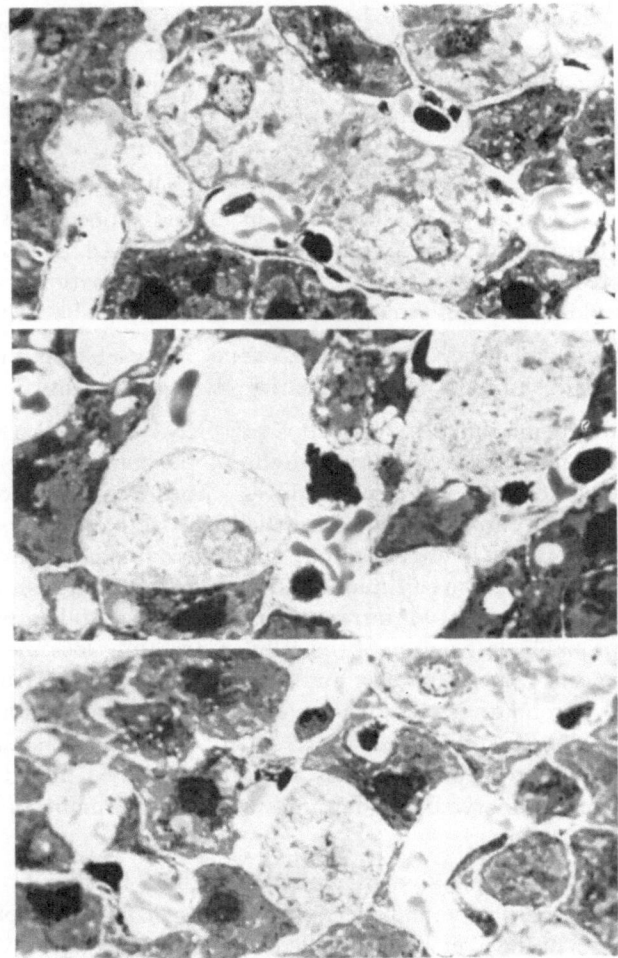

Fig. 11 a, b, c. Liver of guinea pig, 3 days after infection (slices of 0.25 μ). a) Glykogen
free liver cells with hydratization of cytoplasm and edema of the nuclei; b) Liver cells with
destroyed cytoplasm; c) Necrotic and denucleated liver cell like "Councilmanbody"

amid the dark silver-grain loaded parenchyma. Still later, increased activity of
mesenchymal cells surrounding the necrotic areas takes place (Fig. 13a, b).

The electron microscope survey of ultra-thin slices taken from the same blocks
as the semi-thin slices revealed findings which are of great importance for the
interpretation of microscopically demonstrated cellular alterations. 4 days after
the infection one can find clusters of liver epithelial cells which the electron
microscope shows as containing sharply separated areas with focal destruction of

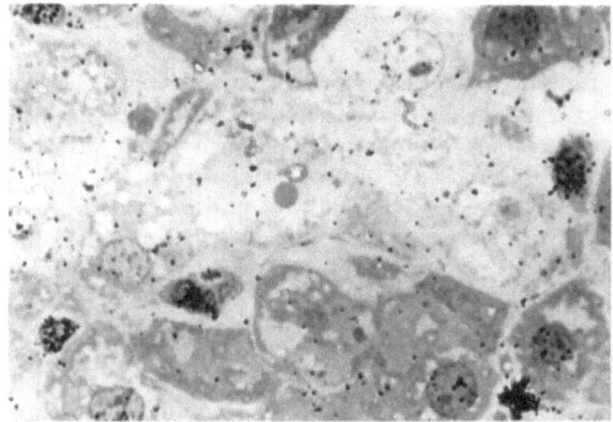

Fig. 12. Autoradiographic detection of the incorporation of Cytidine-3-H in the RNA of parenchymal and mesenchymal cells of the guinea pig liver 3 days after infection. High RNA turnover with high silver grain density in the Kupffer-cells, a lower incorporation in swollen liver cells

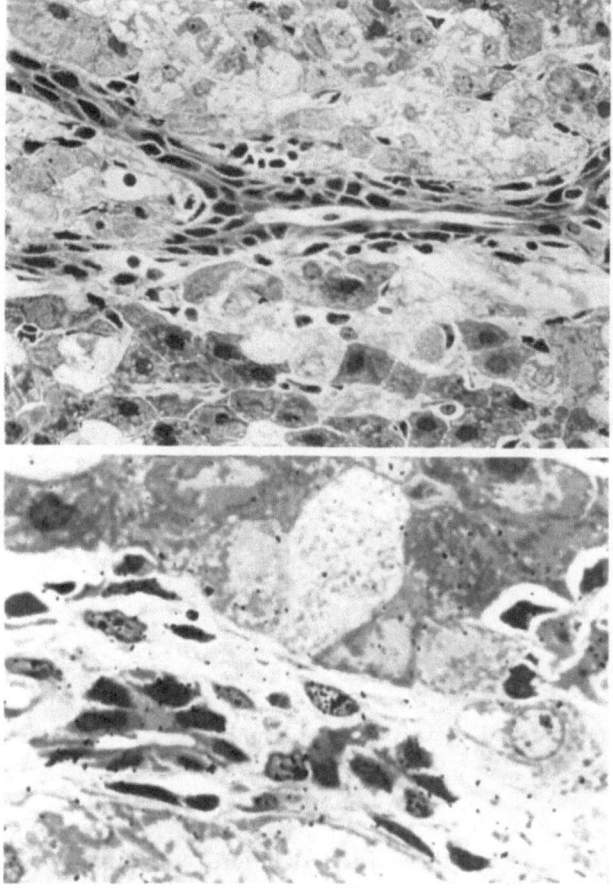

Fig. 13a, b.   Thin slices of guinea pig liver 3 days after infection.   a) Activation of mesenchymal cells in the neighbourhood of swollen and glykogen free liver cells; b) A high incorporation rate of Cytidine-3H in the RNA of activated mesenchymal cells

all cellular structures (Fig. 14). In their intracellular detritus a netlike structureless material of low density can be recognized. In addition to these alterations one can find areas where only the basic cytoplasm is intact with complete loss of

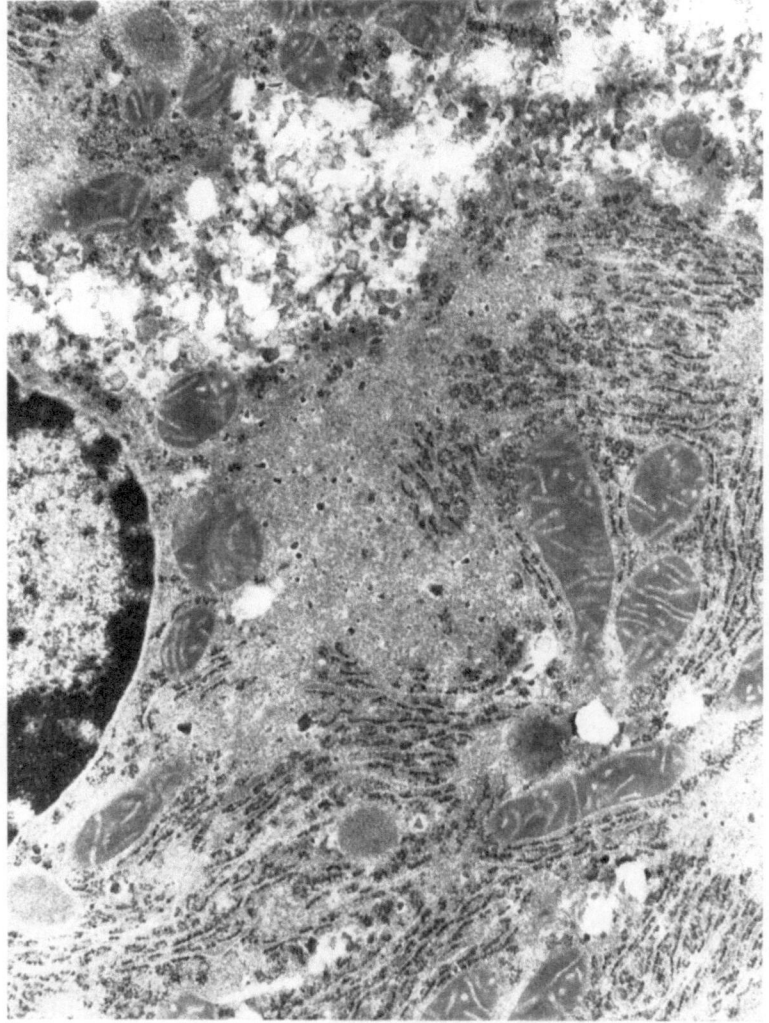

Fig. 14. Homogenized and destroyed cytoplasmic areas in a guinea pig liver cell, 3 days after infection

cellular organelles. Within these areas structures are present which look like ribosomes. Other equally dense, ovally shaped structures of a size up to 1000 Å can be observed. Frequently these structures are surrounded by a light halo at the borderline between these granulated and the completely destroyed cytoplasmatic areas. These generally oval-shaped, dense structures change to small vesicles filled with granules measuring about 50 Å, the size of the vesicles themselves being about 1000—2000 Å. Unquestionably the larger structures found by us, especially those with a lighter halo, look like those structures in the liver and serum of guinea pigs which have been described as viruses by MÜLLER and PETERS [5]. But possibly these structures derive from the debris of decayed endoplasmatic reticulum. The structures surrounded by a halo can also be demonstrated in the liver sinusoids. The tissue alterations in the spleen and lymph nodes of guinea pigs looked like those found in monkeys. Necroses of the follicles and decaying cells of red pulpa were abundant. As seen in the monkeys, we recognized debris of nuclei and accumulation of fibrin with polymorphonuclear leucocytes and red blood cells. After the fifth day of infection a change in the histologic pattern ensued in spleen and lymph nodes. Massive cellular proliferation takes place and is documented autoradiographically by excessive uptake of cytidin. Macronuclear cells, which cannot be differentiated further and which may be called blasts, predominate. These cells revealing numerous mitoses sporadically showed a tender granulation of cytoplasm, suggesting that they represent myelocytes or promyelocytes. Further clusters of micronuclear cells with scanty, deeply eosinophilic cytoplasm were demonstrated in the spleen parenchyma; these are probably identical with erythroblasts. Finally, some polynuclear giant cells representing megakaryocytes are encountered. Until the seventh day after infection the cellular chemistry and the number of megakaryocytes continuously increase, as do monocytic cellular elements; this findings has been stressed by GEDIGK and coworkers [7] during the course of infection in humans.

Summary

The morphological picture and the time sequence of the development of the reactions in liver and spleen of monkeys and guinea pigs after infection with the "Vervet agent" is described.

The first changes seen in the monkey liver are round cytoplasmic inclusion bodies, which appear 3 days after infection with human blood. All the monkeys died 7 to 9 days after intramuscular inoculation of infectious blood. In the liver are found necroses with surrounding hemorrhage, but without inflammation or proliferative reaction of the mesenchymal cells. The cells of the reticulo-endothelial system in liver, spleen, and lymph nodes are destroyed.

Autoradiographic and electron microscope investigations showed in the liver cells of guinea pigs 3 days after infection a loss of cytoplasmic glycogen and focal necrosis of cytoplasmic areas. In the early stages of infection the cells of RES are activated with a high RNA turnover and show proliferative activity. In spleen and lymph nodes of guinea pigs we found 5 to 7 days after infection, in addition to the destruction of reticulum cells in the red pulp and necrosis of germ centers, a high rate of proliferation of mononuclear and multinuclear giant cells.

References

1. Haas, R., Maass, G., Oehlert, W.: Untersuchungen zur Tierpathogenität eines von Cercopithecus aethiops übertragenen menschenpathogenen Erregers. Med. Klin. **35,** 1359 (1968).
2. — — —: Experimental infection of M.rhesus and C. aethiops with the "Vervet agens". Europ. Symp. of the Use of Non Human Primates in Med. Res. Lyon, 1967.
3. — —, Müller, J., Oehlert, W.: Experimentelle Infektionen von Cercopithecus aethiops mit dem Erreger des Frankfurt-Marburg-Syndroms. Z. med. Mikrobiol. u. Immunologie **154,** 210 (1968).
4. Gordon Smith, C. E., Simpson, D. I. H., Bowen, E. T. W., Zlotnik T.: Fatal human disease from vervet monkey. Lancet XX 1967, 1119.
5. Peters, D., Müller, G.: Die elektronenmikrospopische Erkennung und Charakterisierung des Marburger Erregers. Dtsch. Ärzteblatt **34,** 1831 (1968).
6. Siegert, R., Shu, H. L., Slenczka, W.: Isolierung und Indentifizierung des „Marburg-Virus". Dtsch. Med. Wschr. **93,** 604 (1968).
7. Gedigk, P., Bechtelsheimer, H., Korb, G.: Die pathologische Anatomie der Marburg-Virus-Krankheit. Dtsch. Med. Wschr. **93,** 590 (1968).

Laboratory Diagnosis and Pathogenesis

R. Siegert and W. Slenczka

After having clarified the etiology [1], we directed our efforts towards a systematic study of establishing a diagnosis [2, 3]. Let us first discuss the isolation of the agent. For this purpose deep-frozen material from the acute phase of the disease was available, namely 17 blood, 6 throat-washings, 4 urine, 5 stool, and 6 liver specimens. The Marburg virus was found in all investigated blood and liver specimens from the febrile phase. These materials contained the highest concentrations of both agent and antigen, thus rendering the demonstration of the virus highly successful, as has also been shown by others [4, 5, 6].

The probability of isolating the virus from throat-washings and urine is considerably lower. With these materials we were successful twice and once, respectively. The concentrations were at the borderline of detectability. Our efforts were unsuccessful with 5 stool specimens. Thus, excreta are not suitable for the diagnostic isolation of this agent.

On the basis of our experience we recommend for future use the following procedure, which has been shown to be of value with our most recent patients.

With such a dangerous disease, clearing up the etiology within the shortest possible time is mandatory. Fortunately, the high concentrations of the virus in blood and organs make a quick diagnosis feasible. Our method consists of direct demonstration of the antigen by means of immunofluorescence and direct demonstration of the virus using the electron microscope. If drops of blood are examined, the antigen may be found extracellularly in clumps of $1-10\ \mu$. Similar structures may be seen in smears from liver biopsies. They may give the impression of extruded inclusion bodies. These findings, obtained in our last patient, were confirmed in animal experiments.

The second possibility of making a rapid diagnosis lies in the direct demonstration of the virus with the electron microscope. This may succeed if the agent is directly centrifuged from serum or plasma on a carrier, according to Peters and Müller [7]. The characteristic morphology leaves no doubt as to the diagnosis. However, we have no experience enabling us to answer the question concerning the probability that the agent may be overlooked with this method. It is for this reason that attempts to isolate the virus in guinea pigs and in various cell culture systems should follow in any case. If, after the intraperitoneal inoculation of the material, fever develops within ten days, its specificity must be established. The easiest way to do this is by challenging the surviving animals and/or by demonstration of antibodies. Furthermore, passages in guinea pigs should be performed to permit visualization of specific immunofluorescing inclusion bodies in smears from the liver.

In cell cultures cytopathic alterations are absent. However, as early as the third day cytoplasmic antigen inclusions may be recognized, which increase considerably in number and size until the 7th day. They are visible not only by immunofluorescence but also by phase contrast microscopy and in the light microscope after having been stained according to SELLERS.

It may be assumed from the investigations of SIMPSON et al. [8] and HAAS et al. [9] that these suggestions are also applicable to the monkey.

In contrast to the detection of the virus, the problems encountered in attempting a serological diagnosis have not yet been satisfactorily solved. The neutralization test which was originally used by us is much too cumbersome and time-consuming, although titers of up to 1 : 256 could be demonstrated; these are higher than the complement-fixing antibody titers which had been found heretofore. Non-specific virus-inhibiting substances were not encountered in the sera. Because of the low cytopathogenicity of the Marburg virus, the neutralization test in cell cultures is also unsatisfactory. For this reason SLENCZKA [10] has assayed the neutralization capacity of sera by comparing the numbers of inclusion bodies in infected cell cultures with and without added immune sera; this is comparable to a plaque reduction method. However, the results are difficult to reproduce.

Hence for the demonstration of antibodies we rely on the complement fixation reaction. The crude antigen prepared from infected guinea pig livers as used by other teams (SMITH et al. [4]; KISSLING et al. [11]; HOFMANN and KUNZ [12]; KALTER et al.[16]) was unsatisfactory in regard to its specificity. Nor were crude preparations from patient liver of use. Dr. SLENCZKA has already described our efforts to derive better antigens. The antigen from infected Vero cells, which is now in use in our laboratory, is distinguished by its high specificity; serologic crossings of the Marburg agent with other viruses are unknown. Admittedly, we are not yet satisfied with the new antigen's activity; this problem, however, should be overcome by purification and concentration. The increase of the titer of the complement-fixing antibodies occurred within the usual time, i.e. 2—3 weeks after onset. It attained values of 1 : 32, assayed with a nonconcentrated antigen. As a rule, the titers proved constant over many months. Precise data concerning the persistence of antibodies should be available in a few weeks, when our investigations are completed.

As regards the pathogenesis of the disease in man, apparently the virulence of the Marburg virus differs in monkey, man, and guinea pig. Furthermore, an attenuation occurred during the transfer from monkey to man. In the monkeys the experimental infections were always lethal, independent of route and dose [8, 9]. All 7 fatalities in man occurred in people who had acquired the infection directly from monkeys or from cell cultures prepared from these animals. In contrast, the illnesses which were observed after one human passage took lighter, non-fatal courses. In guinea pigs the first passage was practically never accompanied by deaths; however, additional passages led to adaptation which was marked by a rapid increase of lethality. Upon transfer from the blood of the patients or guinea pigs back to the monkeys, the high virulence for the original host had not changed.

The routes by which the agent entered the human body were ascertainable for several patients. They included lesions of the skin, in one case erosions of the mucous membrane of the vagina. Besides injured skin and mucous membranes, the conjunctiva and the upper respiratory organs, but not, however, the gastro-intestinal tract, should be considered possible routes of entry, provided the experiments with guinea pigs may be assumed to offer a clue. Vector transmission by *Aedes aegypti* appears probable (Kunz et al. [13]) but has not been conclusively proved.

The agent then passes into the bloodstream. The viremic phase is characterized by high fever. It lasts approximately 14 days in which time virus concentrations may reach approximately 10^3 guinea pig infectious doses per ml. After the temperature had returned to normal, we never found virus in the blood.

With the blood the virus is carried to all organs and causes necroses, which justifies its characterization as "pantropic". Occasionally it is encountered in throat secretions and urine, which is not surprising, as in many patients enanthema or kidney involvement were in evidence. Stille and his collaborators [14] have drawn attention to the cyclic course of the disease, which either leads to death or to recovery of the patient. Apparently all infections led to manifest illnesses; so far, no serological evidence for an inapparent infection has come to our attention.

Of particular pathogenetic interest are virus carriers whose infection persists in spite of circulating antibodies. Such latent infection was demonstrated in a man who excreted the virus in his sperm. Nothing is known of the further course in this patient for he became impotent, either as a consequence of the infection or for psychological reasons. Nine more men who were checked at the same time did not excrete the virus.

Further carriers may be seen in those cases which developed and endogenous recurrences (Martini and coworkers [15]). Two cases of hepatitis and one of orchitis were observed. They developed between the 31st and the 80th day after onset of the disease. In one of these patients with hepatitis, demonstration of the virus succeeded in a liver biopsy, while the blood specimen, taken at the same time, was free. Hence the recurrence may be assumed to be limited to one organ in which the virus persists. Recrudescence in form of a systemic disease should not be expected because of the presence of neutralizing antibodies.

In two children, born $1\frac{1}{2}$ years after the mother's illness no infectious virus was found in placentae or blood; development was normal. A search for antigen and antibodies is in progress.

It is not yet possible to draw conclusions concerning the consequences of persistence of the virus. Nor can we say how long immunity is likely to last. Therefore the patients must be kept under observation.

References

1. Siegert, R., Shu, H. L., Slenczka, W., Peters, D., Müller, G.: Dtsch. med. Wschr. **92**, 2341–2343 (1967).
2. Siegert, R., Shu, H. L., Slenczka, W.: Dtsch. med. Wschr. **93**, 604–612 (1968).
3. Siegert, R., Shu, H. L., Slenczka, W.: Dtsch. med. Wschr. **93**, 616–619 (1968).
4. Smith, C. E. G., Simpson, D. I. H., Bowen, E. T. W., Zlotnik, I.: Lancet **1967, 2** 1119–1121.
5. Kunz, Ch., Hofmann, H., Kovac, W., Stockinger, L.: Wien. klin. Wschr. **80**, 161–162 (1968).

160   R. Siegert and W. Slenczka

6. May, G., Knothe, H.: Dtsch. med. Wschr. **93,** 620—622 (1968).
7. Peters, D., Müller, G.: D. Ä. **65,** 1831—1834 (1968).
8. Simpson, D. I. H., Bowen, E. T. W., Bright, W. F.: Lab. Anim. **2,** 75—81 (1968).
9. Haas, R., Maass, G., Oehlert, W.: Med. Klin. **63,** 1359—1363 (1968).
   Haas, R., Maass, G., Müller, J., Oehlert, W.: Z. med. Mikrobiol. u. Immunol. **154,** 210—220 (1968).
10. Slenczka, W.: Vortr. 2. Arbeitstagung Dtsch. Ges. Hyg. u. Mikrobiol. in Mainz 8. 10. 1968, Zbl. Bakt. I Ref. **215,** 545—546 (1969).
11. Kissling, R. E., Robinson, R. Q., Murphy, F. A., Whitfield, S. G.: Science **160,** 888—890 (1968).
12. Hofmann, H., Kunz, Ch.: Zbl. Bakt. I Orig. **209,** 288—293 (1969).
13. Kunz, Ch, Hofmann, H., Aspöck, H.: Zbl. Bakt. I Orig. **208,** 347—349 (1968).
14. Stille, W., Böhle, E., Helm, E., van Rey, W., Siede, W.: Dtsch. med. Wschr. **93,** 572—582 (1968).
15. Martini, G. A., Knauff, H. G., Schmidt, H. A., Mayer, G., Baltzer, G.: Dtsch. med. Wschr. **93,** 559—571 (1968).
16. Kalter, S. S., Ratner, J. J., Heberling, R. L.: Proc. Soc. Exptl. Biol. Med. **130,** 10—12 (1969).

Epidemiology of "Marburg Virus" Disease

W. HENNESSEN

"In August 1967 a number of persons in Germany who had contact with the blood or organs of a single batch of vervet monkeys became ill with a hemorrhagic disease. Further cases occurred in September in Belgrade." Seven of the thirty cases were fatal.

This quotation is the first sentence from the report of an ad hoc committee set up by the Permanent Section of Microbiological Standardization in 1967 published in the National Cancer Institute Monograph 29, 1968.

In the following paper an attempt is made to describe some observations which seem necessary to understand the epidemiology of the disease.

Human Cases

The distribution of the cases for the three affected institutions is given in Table 1.

Table 1.

		Cases	Fatal	Secondary
Marburg		20	5	2
Frankfurt		4	2	2
Belgrade		1	—	1
	Total	25	7	5

The majority of cases occurred in Marburg, the number of monkeys used at that time was largest there too.

The sequence of events can be seen from Table 2.

The dates of the total number given on the upper part show the explosive nature of the infection in man. This virulence is underlined by the fact, that from 25 primary cases 5 additional secondary cases contracted the disease, an apparently unusually high percentage.

The occupational hazards of the patients can be divided into the four groups shown in Table 3.

Table 2.

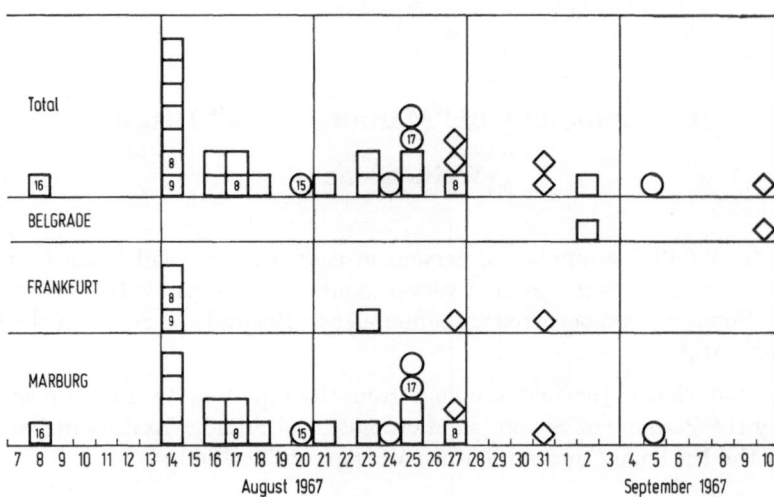

Distribution of cases

☐   Persons performing the surgical procedures
○   Persons working with monkey organs outside animal quarters
◇   Secondary cases

Table 3.

	Group	Patients	Fatalities
1	Surgical procedures with monkeys, including removal of kidneys and brain, opening of thoracic and abdominal cavities, autopsies, and taking blood samples from the monkeys inside their quarters	20	5
2	Work with monkey organs outside animal quarters, such as trypsinization and mincing of kidneys	1	1
3	Cleaning of tissue culture containers	3	1
4	Accident, broken tube	1	0

The actual occupational risk may be seen from the fact that 20 of the 29 persons with blood-contact fell ill while 4 of the 13 exposed to tissue cultures were infected.

It should be mentioned here that animal attendants who were not in contact with blood were not infected, nor were the personnel who took the usual precautions when working with viruses.

Incubation time

The epidemiological analysis of the 20 primary cases in Marburg is based on the knowledge of the incubation period in man and in monkeys. For primary infections an incubation time from 3—7 days was found, while secondary cases occurred 5—8 days after exposure. Parenteral infection of monkeys with infectious material from man resulted in a disease with high temperatures after 3—4 days, while monkeys experimentally exposed to infected animals in neighboring cages contracted the disease after 6—9 days.

Epidemiological Analysis

Knowing the possible time of exposure and working conditions of the patients, one can postulate the minimal number of foci of infection. Since the Marburg monkeys were kept two in a cage, one source of infection there probably means two monkeys.

Table 4. *Analysis of Time of Contact*

Patient No.	Age (years)	Most probable contact with shipment of:		focus of infection
1	28	28. 7.		I
2	41	28. 7.		
3	30	28. 7.		
4	32	28. 7.		
5	21	28. 7.		II
6	52	28. 7.		
7	64	28. 7.		
8	39	28. 7.		
9	38	28. 7.		
10	20	28. 7.		
11	29	28. 7.		II
12	52	28. 7.		
13	36	28. 7.		
14	58		21. 7.	
15	26		21. 7.	
16	19		21. 7.	IV
17	30		21. 7.	
18	20		21. 7.	
19	40		21. 7.	V
20	24		21. 7.	(IV)

As may be seen from Table 4, for Marburg we have to consider at least five different foci of infection to explain all 20 cases; this means that a minimum of 10 monkeys was infectious during the time they were worked on by the patients. For the cases in Frankfurt another two foci—there this may mean one animal each—and for the case at Belgrade another monkey have to be regarded as infectious. From such considerations we arrive at a minimal number of 13 infectious animals. They originated from 4 shipments reaching Europe from Uganda between July 20 and August 10. The total number of animals in these shipments was between 500 and 600. The distribution of these animals to the three affected institutions varied greatly. Frankfurt, for instance, received less than 10% of the total and this small number was divided into two shipments. With an infectivity rate of roughly 2%, the likelihood that two infective monkeys should have reached Frankfurt on two occasions is remote.

Our minimal calculation may permit a conclusion of some epidemiological significance: it is unlikely that all animals who were the minimal foci of infection for man reached the three institutions in the infective state. If this is true, there must either have been more monkeys infected from the beginning, or infection spread within these institutions. In the first case, one would have expected an elevated mortality among these animals. This did not occur. The latter assumption seems far more likely, which means that there was a spread of infection within the animal quarters.

The mode of spread is difficult to determine. In two of the three affected places it may have been both airborne and direct transmission. In Marburg, however, the monkeys were housed in separate rooms for every shipment under an airconditioning system with separate in- and outlets for every room and no recirculation. Here transmission by air seems highly unlikely, although infection via mobile equipment cannot be excluded and may have occurred. It may be mentioned here that the spread from one shipment to another took place within a maximum period of three weeks.

Origin of Infection

Our observations have hitherto been limited to events and conclusions from events that occurred within the three affected places. Keeping in mind all shortcomings of extrapolation in epidemiology, one may feel tempted to reconstruct what must have happened to these monkeys before they reached their destination in continental Europe. It could be shown experimentally that infected monkeys become infectious themselves from the 5th to the 8th day post infect. Assuming this minimum period, the monkey who infected the first, the index case in Marburg, must have been infected himself within 4 days outside continental Europe. Of these 4 days he spent 60—87 hours at the farm in Uganda and on board the cargo plane, and during the remaining 9—36 hours he was in the custody of the RSPCA hostel at London airport.

The PSMSt-Report on this transit reads: "During this time there were no less than 48 species of animals and birds from all over the world that could have had indirect contact with the vervets. In particular, 1 of 2 langurs (Presbytis) being shipped from a Ceylon zoo to Holland, became ill and died on August 1. It may also be significant that the monkeys had the greatest contact with a large variety of birds from many parts of the world; there were also rhesus monkeys in a different room in the hostel."

Control Measures

Measures taken to stop new cases were guided by the thought that all monkeys, including their organs and blood, might be infectious regardless of the time of quarantine. Measures were carried out in three steps within one week. The first step included compulsory wearing of protective clothing with gloves and masks not only during work with monkeys—that had been done before—but also for cleaning up after this work etc. In addition, disinfectants were used for all cleaning and washing procedures (18. 8.). The second step terminated all tissue culture work from monkey organs (22. 8.). Finally, the remaining monkeys were sacrificed and sterilized in the incinerator (25. 8.).

Judging from the further occurrence of cases, the first step proved already successful, because new cases ceased within one incubation period. The two last cases could not possibly have been prevented by these measures. One occurred in the veterinary surgeon in charge of the monkey quarters. He insisted on doing autopsies of monkeys without any help while the preliminary diagnosis of the disease in man was still Shigella (21. + 22. 8.). The last case was in a person who entered the sterilization area without permission. She stated that she broke a tissue culture vessel taken from infected material which was to be sterilized (1. 9.).

Conclusions

The observations mentioned above seem to permit the following statement on the incident:

An agent infective for monkey and man was introduced by infected monkeys into the animal quarters of three institutions. Not more than 2—3 originally infective animals were necessary as foci to explain the whole episode, since infection among monkeys spread through vectors which were most probably parts of the equipment. There is no indication of transmission by air.

Infection from monkey to man occurred through direct contact with blood or organs of the animals including tissue culture. Airborne transmission to man could be excluded. The origin of the infective agent could be traced by theoretical estimation to foci outside continental Europe.

Measures breaking direct contact, especially contact with contaminated equipment during cleaning and washing, stopped the occurrence of new cases within one incubation period.

References

1. HAAS, R., MAASS, G., OEHLERT, W.: III. Experimental Infections of Monkeys. Primates in Med., **3,** 138—139 (Basel/New York: Karger, 1969).
2. HENNESSEN, W.: A Hemorrhagic Disease Transmitted from Monkeys to Man. US Nat. Canc. Inst. Monogr. **29,** 161—171 (1968).
3. HENNESSEN, W., BONIN, O., MAULER, R.: Zur Epidemiologie der Erkrankung von Menschen durch Affen. Dtsch. Med. Wschr. **93,** 582—589 (1968).
4. SIEGERT, R., SHU, HSIN-LU, SLENCZKA, W., PETERS, D., MUELLER, G.: Zur Ätiologie einer unbekannten, von Affen ausgegangenen menschlichen Infektionskrankheit. Dtsch. Med. Wschr. **92,** 2341—2343 (1967).
5. SIEGERT, R., SHU, H.-L., SLENCZKA, W.: Isolierung und Identifizierung des „Marburg-Virus". Dtsch. Med. Wschr. **93,** 604—612 (1968).
6. SMITH, C. E. GORDON, SIMPSON, D. I. H., BOWEN, E. T. W., ZLOTNIK, I.: Experimental Production of Microangiopathic haemolytic Anaemia in Vivo. Lancet **1967 I,** 1119—1121.

Epidemiological Studies in Uganda Relating to the "Marburg" Agent

B. E. Henderson, R. E. Kissling, M. C. Williams, G. W. Kafuko, and
M. Martin

With 1 Figure

Introduction

During August and September 1967 several cases of an acute, severe hemorrhagic disease were recognized in laboratory workers in Germany and Yugoslavia [1—3]. Epidemiological studies implicated contact with tissues of African green or vervet monkeys *Cercopithecus aethiops* (L.) as common to all primary cases [4]. The monkeys incriminated were all originally exported from Uganda for use in production and testing of tissue culture-derived vaccines.

Upon receiving news of the outbreak, field investigations were initiated in Uganda to determine if there was evidence of a comparable disease problem in monkeys or humans. Initial inquiry revealed that no unusual incidence of illness had occurred among monkey trappers or among trapped vervet monkeys. Selected specimens were collected for attempted isolation and serologic studies overseas. The studies reported below amplify a preliminary report of work done in Uganda[5] and give new serologic findings recently made at the Center for Disease Control (CDC).

Materials and Methods

Collection of Specimens

Serum or whole blood was collected from primates by femoral venipuncture, using sterile disposable plastic syringes. Livecaptured wild rodents or experimental guinea pigs were bled by intracardiac puncture. Additional specimens including brain, liver, spleen, and kidneys were collected from wild rodents by sterile dissection of anesthetized animals. All specimens were stored at −60 °C until processed.

Shipment of Specimens

Serum specimens were held in Entebbe at −60 °C until shipment by air on dry ice to the Center for Disease Control, Atlanta, Georgia.

Initially, undiluted serum specimens from live captured monkeys were inoculated by three routes into newborn white Swiss mice: 0.01 ml intracerebrally (IC), 0.01 ml intraperitoneally (IP) and 0.01 ml subcutaneously (SC). Isolation attempts were later performed in adult male guinea pigs. Aliquots of 0.5 ml of whole blood from primates were inoculated directly by the IP route into guinea pigs.

Specimens of liver and spleen from wild rodents were ground with mortar and pestle and ground glass and, 10% suspensions were centrifuged at 2500 RPM for 10 minutes. The supernatant was decanted and 0.5 ml was inoculated IP into adult male guinea pigs.

Inoculated mice were observed for a period of three weeks and then bled and discarded. Adult guinea pigs were bled and then baseline rectal temperatures recorded twice daily for at least two days prior to inoculation. After inoculation, rectal temperatures were recorded for 35 consecutive days at which time the animals were bled and discarded. No attempt was made to autopsy or passage specimens from guinea pigs which were sick or found dead. Carcasses of dead guinea pigs were incinerated.

Complement Fixation Tests

Complement fixation (CF) tests were performed by the micro LBCF technique [6]. CF antigen was prepared from the livers of guinea pigs moribund following inoculation with the Porton strain of Marburg virus. Sufficient borate saline (pH 9.0) was added to minced liver in a flask containing glass beads to make a 20% tissue suspension. Beta propiolactone was added to make a final concentration of 1:2,000 to 1:3,000. The flask was stoppered tightly and shaken to effect emulsion of the liver pieces. The emulsion was placed at 4 °C for 3 to 4 days and shaken occasionally. No attempt was made to adjust the pH. The material was then allowed to settle, the supernatant removed and used as antigen without further treatment. The antigen was stored at −60 °C. Normal guinea pig liver processed in a similar manner was used as a control antigen. A 1:10 dilution of antigen, the optimal reaction endpoint, was used for routine testing of sera.

Many of the monkey sera were either anti-complementary or reacted non-specifically with the control antigen when sera was only heat treated (56 °C/30 minutes). Adsorption with hamster liver powder was partially effective in removing some of these non-specific reactions. However, treatment with CO_2 while not quite as efficient proved more practical for large scale serum treatment (Table 1).

Table 1. *Comparison of methods for treating monkey serums to remove non-specific complement fixation reactions*

Treatment of Antigen

56 °C/30' only		Liver Powder		CO_2	
Marburg	Control	Marburg	Control	Marburg	Control
32*	16	16	<8	32	8
128	32	32	16	128	16
16	8	<8	<8	16	<8
128	64	16	<8	8	<8
N. T.	N. T.	8	<8	32	8
N. T.	N. T.	<8	<8	8	8
N. T.	N. T.	32	<8	128	<8
N. T.	N. T.	32	<8	64	8
N. T.	N. T.	8	<8	32	8

* Reciprocal of serum titer.

CO_2 treatment consisted of preparing a 1:7 dilution of the serum in distilled water and adding a small piece of dry ice. The precipitate which formed was removed by centrifugation. The isotonicity of the supernatant fluid was restored by addition of NaCl; the preparation was then heated at 56 °C/30 minutes,

A serum was not interpreted as positive by the CF test unless the titer observed against Marburg antigen was at least 4-fold higher than the reaction obtained with control antigen. A titer of 1 : 8 with monkey sera was not considered positive since this was the lowest dilution tested.

Neutralization Tests

Neutralization tests (NT) were performed in tube cultures of the AH-1 line of green monkey kidney cells. Virus used in the NT was the Porton strain which had been adapted to BHK 21 cells. Sera were initially diluted to 1 : 4 or 1 : 8 in Hank's BSS and heated at 56 °C for 30 minutes. Sera were then diluted serially in 2-fold steps and an equal volume of virus dilution containing 30 to 300 50% tissue culture infective doses ($TCID_{50}$) per 0.1 ml was added. The serum-virus mixtures were incubated at 37 °C for 1 hour before inoculating the tissue cultures with 0.2 ml of the mixture. Endpoints were determined by observation of cytopathic effect (CPE).

Background Information

Collection of live vervet monkeys for export from Uganda has been an established trade since 1962. Vervet monkeys during 1967 were captured mainly in the area of Lake Kyoga. They were caught in wire cages with food bait and brought to three holding stations (Fig. 1), one at Namasale and two near Kidera. At the

Fig. 1. Map of Uganda showing **1** field holding stations near Lake Kyoga, **2** main holding area at Entebbe, and **3** sites of collection of primates during survey of 1968

holding stations, they were held in individual cages as much as possible although very small monkeys were often kept in pairs. At varying intervals, up to 2 weeks after capture, the monkeys were brought from the collecting station to the central holding area at Entebbe. They were transported in individual compartments and caged singly in Entebbe. After a further holding period of at least 3 days, the monkeys were exported by air to different parts of the world. The number of vervet monkeys exported increased during 1967 and the time between trapping, shipping and experimental use diminished.

During July and August 1967, a total of 1772 vervet monkeys were compounded in Entebbe, the vast majority of these coming from the Lake Kyoga area (Table 2). During the same two-month period of time, 1290 vervet monkeys were exported

Table 2. *Original location of capture of vervet monkeys received at Entebbe holding station from July — August, 1967*

Location	July 1—14	15—28	August 29—11	12—25	Total
Kidera No. 1	378	302	481	179	1340
Kidera No. 2	92	63	46	111	312
Namasale	—	57	—	—	57
Entebbe	7	10	4	17	38
Sesse Is.	—	—	—	25	25
	477	432	531	332	1772

from Uganda, the largest number going to Germany and Yugoslavia (Table 3). Although normally monkeys were transported directly to Germany and Yugoslavia, during July and August some of the monkeys were shipped via London

Table 3. *Country of destination of vervet monkeys exported from Entebbe from July — August, 1967*

Country of Destination	July 1—14	15—28	August 29—11	12—25	Total
Germany	130	250	250	135	765
Japan	0	30	15	35	80
Sweden	30	15	15	25	85
Yugoslavia	0	200	100	—	300
Other*	3	25	20	12	60
Total	163	520	400	207	1290

* Includes Holland, Switzerland, Italy, and Czechoslovakia.

where they were held in animal quarters near the airport until further shipment to the country of destination. While being held at London airport, the monkeys from Uganda were in potential contact with a large variety of animals from many parts of the world.

Careful inquiry revealed no history or unusual illness or death among monkeys either in their natural environment or after capture and transport to Entebbe.

Results

Isolation Attempts (Table 4)

On August 30, 1967, specimens of venous blood were taken from 30 vervet monkeys housed at the Entebbe holding station. These monkeys had arrived at Entebbe during the preceding 2—3 weeks from the Lake Kyoga area. The blood was allowed to clot and the serum separated and inoculated into newborn mice. No illness was observed in the mice.

Table 4. *Material collected for attempted isolation in guinea pigs from area of Lake Kyoga and Entebbe, 1967*

Location	Material Inoculated	No.	Date, Inoc.	No. Surv.	Date Second Bleed.	CF Conv. No. Pos/ No. Tested
Kidera No. 1	WB* *C. aethiops*	22	25/9	20	31/10	0/20
	WB Human control	5	25/9	3	31/10	0/3
Kidera No. 2	WB *C. aethiops*	26	25/9	15	31/10	0/15
	WB Human control	5	25/9	1	31/10	0/1
Namasale	WB *C. aethiops*	19	27/9	15	2/11	0/15
	WB Human control	5	27/9	4	2/11	0/4
Entebbe	S** *C. aethiops*	30	30/9	25	6/11	0/25
	S Guinea pig (control)	6	30/9	5	6/11	0/5
Entebbe	LS*** Rodents	21	2/11	20	6/12	0/18
Kidera No. 2	LS Rodents	13	21/11	13	28/12	0/12
Total		152		121		0/118

  * WB — whole blood
 ** S — Serum
*** LS — 10% suspension of pooled liver and spleen.

When information was received of the successful isolation of an agent from human tissues inoculated into guinea pigs [1], isolation attempts were shifted to this animal system. Monkeys were bled at the three holding stations in the Lake Kyoga area between September 25—27. Whole blood was removed and inoculated directly by the IP route into guinea pigs brought to the field. For each group of 4—5 guinea pigs inoculated with monkey blood, one guinea pig was inoculated with blood from a human volunteer known to be in good health. In addition, the serums of the 30 monkeys originally bled in Entebbe and liver-spleen suspensions of 44 rodents captured at Entebbe and at the field stations were inoculated into guinea pigs. None of the inoculated guinea pigs showed a sustained febrile response. An apparent outbreak of gastroenteritis was noticed in the guinea pig holding room during mid-October and was associated with mortality in both experimental and control guinea pigs. Acute and 35-day convalescent serum were tested on 118 guinea pigs and none of these sera had detectable CF antibody to the CDC Marburg antigen.

Serology

Monkeys. CF tests were performed on the serums of 49 *C. aethiops* monkeys bled between August 30 and September 18, 1967, at the Entebbe holding station. The serum from only one of these monkeys reacted at a titer of >1: 8 (Table 5).

Table 5. *Number of sera from vervet monkeys with CF antibody*

Location	Aug. 30 – Sept. 30*			Oct. 1 – 16		
	Tested	Pos**	% Pos	Tested	Pos	% Pos
Entebbe	49	1	2	33	11	33
Kidera	49	10	20	33	12	36
Namasale	11	0	0	11	1	9
Total	109	11	10	77	24	31

* Monkeys from Entebbe bled between August 30 – September 15.
  Monkeys from Kidera and Namasale bled between September 15 – 30.
** CF titer ≥1 : 16.

This monkey was bled on September 13 and had recently been transported from the Lake Kyoga area to Entebbe. During late September 60 *C. aethiops* were bled at the three Lake Kyoga holding stations and the sera from 10 *C. aethiops* had titers of >1: 8.

In the first half of October a second blood sample was collected from 38 of the 109 *C. aethiops* sampled earlier at Entebbe or near Lake Kyoga. In addition specimens were obtained from 39 *C. aethiops* not tested earlier. CF titers of >1: 8 were detected in the serums of 24 of these 77 monkeys.

Table 6. *Results of CF tests on monkeys from Kidera and Entebbe showing serological conversion between paired specimens*

Location	Monkey	Acute		Convalescent	
		Date	Titer	Date	Titer
Kidera	4	25/9	<8	27/10	16
	17	25/9	<8	27/10	128
	20	25/9	<8	27/10	16
	25	25/9	<8	27/10	16
	28	25/9	8	27/10	32
Entebbe	57	4/9	<8	28/10	16
	86	6/9	<8	28/10	64
	89	6/9	<8	28/10	32
	135	13/9	32	28/10	128

Of the 38 monkeys with paired specimens 21 had CF titers ≤1: 8 in both specimens and 8 had CF titers ≥1: 16 in both specimens with no more than a two-fold difference in titer. The remaining nine monkeys had a 4-fold or greater rise in CF antibody titer (Table 6).

During 1968, additional monkey serum specimens were obtained from several areas in Uganda including a new sample of monkeys from the Lake Kyoga area (Table 7). The overall incidence of positive sera was 10%. One of six redtail monkeys (*Cercopithecus ascanius schmidti* Matschie) from Masaka was also positive.

Table 7. *Results of CF tests on sera collected from monkeys in several regions of Uganda during 1968*

Location	Species	Age*	No. Tested	No. CF** Positive
Masaka	*C. ascanius*	A	6	1
	C. aethiops	A	2	0
	C. aethiops	J	7	0
Karamoja	*C. aethiops*	A	7	0
	C. aethiops	J	4	0
West Mengo	*C. aethiops*	A	6	2
Kidera	*C. aethiops*	A	45	5
Laropi	*C. aethiops*	A	60	4
Total			164	16

 * A = adults, J = juveniles under 18 months of age.
** CF titer > 1 : 8.

A limited number of neutralization tests were performed with serums from *C. aethiops*. Three vervet sera negative in CF tests contained no demonstrable neutralizing antibody (Table 8). Three serums giving positive CF reactions also

Table 8. *Comparison of CF and NT antibody titers in serums of vervet monkeys*

Serum No.	CF Titer	NT Titer
9527	<8	< 8
9540	<8	< 8
12690	<8	< 4
9041	32	≥32
10818	32	8
13057	8	16
9506	<8	16
13038	<8	16

had positive neutralization titers. Two additional serums with negative CF reactions were capable of neutralizing Marburg virus. One of these latter serums was from a monkey bled in 1968, one year after the period of apparent increased virus activity in the Uganda monkeys. Additional neutralization tests were performed on serums of three monkeys from which paired specimens had been taken (Table 9). One animal had no demonstrable neutralizing antibody in either specimen, although CF antibody titer increased from negative in the first specimen to 1 : 16 in the second. A second animal had very low neutralizing antibody in both specimens

(1: 4—1: 8), while the CF antibody titers increased from less than 1: 8 in the first specimen to 1: 32 in the second. A third monkey had CF antibody in both specimens, increasing from 1: 32 to 1: 128, indicating its infection had probably occur-

Table 9. *Comparison of CF and NT titers in serially bled Ugandan vervets*

Monkey No.	Test	Titer* Serum No. 1	Serum No. 2
57	CF	<8	16
	NT	<8 (Sept. 4)	< 4 (Oct. 28)
89	CF	<8	32
	NT	8 (Sept. 6)	4 (Oct. 28)
135	CF	32	128
	NT	NT (Sept. 13)	≥128 (Oct. 28)

* Reciprocal

red earlier than in the cases of the other two animals. Although an insufficient amount of the first serum was available for a neutralization test, the second specimen had a neutralizing titer of ≥1: 128.

Other Wild Animals. Sera from nine bush babies (*Galago senegelensis*) were negative by CF. These specimens were collected at Lake Kyoga in the vicinity of trapping stations near Kidera.

At both the holding stations near Lake Kyoga and at Entebbe, large numbers of rodents were observed in the vicinity of the monkey cages. For this reason sera were collected from 24 rodents at Entebbe and 56 rodents at Lake Kyoga. Among the peridomestic and woodland species captured were *Arvicanthis abyssinicus* (RUPPELL), *Hylomyscus stella* (THOMAS), *Lophuromys flavopunctatus* (THOMAS), *Mastomys natalensis* (SMITH), and *Praomys morio* (TROUESSART). Only one serum, collected from a male *A. abyssinicus*, was positive with a CF titer of 1: 16.

Humans. During August-September, 1967, 79 persons who were engaged in trapping or handling *C. aethiops* monkey at Lake Kyoga or Entebbe were interviewed and bled. None of these persons reported an unusual illness during the

Table 10. *Comparative CF and NT tests on serum from humans trapping monkeys in Uganda*

Serum No.	Age/Sex	Locality	CF Titer	NT Titer
2845	34 M	Entebbe	<8	<8
3057	46 M	Kidera	8	32
3067	43 M	Kidera	32	32

preceding 12 months; however, 3/79 sera had CF titers of ≥1: 8. Neutralization tests were performed on two of the CF positive sera and one CF negative serum and the results of both tests were similar (Table 10).

Discussion

The outbreaks of febrile, hemorrhagic disease among laboratory workers in Germany during 1967 stimulated intensive efforts by several laboratories to isolate the causative organism. An agent isolated from tissues of fatal cases and named the "Marburg" agent has been shown to be serologically related to the human disease and has been characterized by electron microscopy [1, 7—11].

Because of the association of these laboratory workers with tissues from vervet monkeys imported from Uganda, our studies concentrated on attempts to determine if the agent was active in Uganda primates. The limited attempts at isolation in guinea pigs were apparently unsuccessful. If adequate high security facilities had been available in Entebbe, passaging of tissues from sick guinea pigs might have provided further information.

Serological studies with the crude CF antigen have indicated presence of CF antibody in the serums of vervet monkeys from the Lake Kyoga collecting stations and Entebbe. The first CF positive monkey was detected on September 13 at Entebbe; the serums of 44 monkeys which arrived at Entebbe from Lake Kyoga at an earlier date were all CF negative. By late September, 20% of the monkeys bled in the Lake Kyoga area were CF positive; and of the monkeys rebled at the Entebbe or the Lake Kyoga holding stations in October, nine serum pairs showed a rise in CF titer over the intervening 4 to 6 week period. The appearance of CF antibody in increasing numbers of serums drawn during late September and October suggests that an infectious agent related or similar to the Marburg virus was circulating in primates in the Lake Kyoga area somewhat earlier, probably during a period extending from July-September.

The number of neutralization tests were limited by the lack of adequate safety facilities at CDC. There is some evidence to suggest that neutralizing antibody rises later than CF antibody. The two monkeys which converted from CF titers of $<1:8$ to $\geq 1:16$ (Table 9) did not have a corresponding rise in neutralizing antibody titer. The third monkey which already had detectable CF antibody in the first serum had a further CF antibody rise and an equivalent level of NT antibody in the second sample.

That vervet and redtail monkeys collected from elsewhere in Uganda have a low incidence of CF antibody suggests that the "Marburg" agent or a closely related agent is widespread in Uganda monkeys. Recently, KALTER [12] has demonstrated CF antibody in a variety of African simians using the same antigen as described in this report. These simians were collected from a variety of areas in Africa and included baboons, (*Papio* sp.), Talapoin (*Cercopithecus talapoin*), vervet (*Cercopithecus aethiops*), chimpanzees (*Pan* sp.) and patas (*Erythrocebus patas*).

Experimentally infected vervet monkeys have been reported to have a very high mortality [13]. Possibly passage of the agent through different host systems may account for the exhibited pathogenicity. Certainly there was no evidence of increased mortality in vervet monkeys naturally infected in Uganda.

The finding of CF and/or neutralizing antibody in 3 monkey trappers indicates that human infection has occurred in Uganda. There was no report of a disease syndrome similar to that seen in the European laboratory workers in these three individuals or in any of their associates. Additional serological surveys of human populations in Africa are required to determine the distribution and frequency of human infection.

Our findings fit the conclusion that an agent closely related to the "Marburg" agent, but with little if any virulence for monkeys, is present in Uganda and was active during August-September 1961. It is necessary to remember that this conclusion is based mainly upon serological tests. The relationship between the CF and NT results has to the further studied and the specifity of the tests, particularly the CF test, is not known at this time.

Summary

Epidemiological studies were undertaken in Uganda in association with the outbreak of hemorrhagic disease among laboratory workers in Germany and Yugoslavia in 1967. No agent was isolated from the blood of wild-trapped *Cercopithecus aethiops* (L) inoculated into guinea pigs. Using a crude antigen prepared from the livers of guinea pigs infected with Marburg virus, a rise in complement-fixation antibody was demonstrated in the sera of nine *C. aethiops* originally trapped near Lake Kyoga. CF antibodies were also demonstrated in the sera of up to 36% of *C. aethiops*, one *C. ascanius*, one *A. abyssinicus* and three monkey trappers.

Acknowledgements

We are indebted to Mr. Tom Mann, of Entebbe, Uganda who generously assisted in the obtaining of specimens from monkeys he had captured. Mr. Mann graciously supplied information on the source and destination of primates captured for export from Uganda. The technical assistance of Dr. P. M. Tukei, Dr. Jack Woodall, M. Lule and J. Ellice is gratefully acknowledged.

References

1. Smith, C. E. G., Simpson, D. I. H., Bowen, E. T. W., Zlotnik, I.: Fatal Human Disease from Vervet Monkeys. Lancet **1967, 2,** 1119—1121.
2. Martini, G. A., Knauff, H. G., Schmidt, H. A., Mayer, G., Baltzer, G.: An infectious disease presumably transmitted from monkeys (Marburg virus disease). Dt. Med. Wschr. **93,** 559—471 (1968).
3. Stille, W., Bohle, E., Helm, E., van Rey, W., Siede, W.: An infectious disease transmitted from green monkeys. Dt. Med. Wschr. **93,** 572—582 (1968).
4. Hennessen, W., Bonin, O., Mauler, R.: On the epidemiology of a disease transmitted from monkeys to man ("Marburg virus disease"). Dt. Med. Wschr. **93,** 582—589 (1968).
5. Williams, M. C., Henderson, B. E., Tukei, P. M., Ellice, J. M., Lule, M., Ssekubuge, Y.: East African Virus Research Institute Annual Report No. 17 Government Printer, Entebbe, Uganda, p. 43—45, 1968.
6. Casey, H. L.: Adaption of LBCF Method to Micro Technique Public Health Monograph No. 74, P.H.S. Pub. No. 1228, U. S. Government Printing Office, Washington, D. C., 1965.
7. Kissling, R. E., Robinson, R. Q., Murphy, F. A., Whitfield, S. G.: Agent of Disease Contracted from Green Monkeys. Science **160,** 888—890 (1968).

8. SIEGERT, R., SHU, H. L., SLENCZKA, W.: Identification and isolation of the "Marburg virus". Dt. Med. Wschr. **93**, 604—612 (1968).
9. SLENCZKA, W., SHU, H. L., PIEPENBURG, G., SIEGERT, R.: Demonstration of antigen of the "Marburg virus" in organs of infected guinea pigs by immunofluorescence. Dt. Med. Wschr. **93**, 612—616 (1968).
10. SIEGERT, R., SHU, H. L., SLENCZKA, W.: Demonstration of the "Marburg virus" in infected patients. Dt. Med. Wschr. **93** 616—619 (1968).
11. ZLOTNIK, I., SIMPSON, D. I. H., HOWARD, D. M. R.: Structure of the Vervet-monkey-disease agent. Lancet **1968, 2**, 26—28.
12. KALTER, S. S., RATNER, J. J., HEBERLING, R. L.: Antibodies in Primates to the Marburg Virus. Proc. Soc. Exp. Biol. Med. **130**, 10—12 (1969).
13. SIEGERT, R., SHU, H. L., SLENCZKA, W., PETERS, D., MULLER, G.: On the cause of a previously unknown human infection transmitted from monkeys. Dt. Med. Wschr. **92**, 2341—2343 (1967).

A Serological Survey of Primate Sera for Antibody to the Marburg Virus[1]

S. S. KALTER

Little doubt remains about the association of African green monkeys *(Cerco-pithecus aethiops)*[2] with the recent outbreak of human disease among laboratory and hospital personnel in Germany and Yugoslavia. What remains unresolved, however, is the epidemiology of this outbreak. The purpose of this report is to present data obtained from an expanded study on sera obtained from various primate sources other than that previously reported [1]. Included herein are results not only from a larger serum series on animals previously assayed, but from additional primate species as well. For comparison, aliquots from the same baboon and African green monkey sera as tested in both Dr. H. Malherbe's laboratory in South Africa and at Southwest Foundation for Research and Education (SFRE) are included, as are 3 human sera supplied by Dr. R. SIEGERT (Marburg, Germany) and hamster and guinea pig sera supplied by Dr. D. I. H. SIMPSON (Porton, England).

Sera from a broad range of primates have been derived from numerous sources long involved in handling primates but without any history of a similar outbreak. These sera are submitted to us for a number of reasons, but primarily to assist our program as World Health Organization (WHO) and National Institutes of Health (NIH) Simian Virus Reference Center. However, with the exception of our facility in San Antonio, little information is available concerning the history of these animals from the time of capture. Again, with the exception of the SFRE animals, it must be assumed that these various simians have experienced frequent, if not constant contact, with one or more primates (including man) of other species and geographic origin. Thus, one of the most important epidemiologic factors, that is specific relationships of host and infectious agent, is very probably lost as a result of this contact.

Materials and Methods

Sera: Sera used in this study were obtained from the following primate species: human *(Homo sapiens)*; gorilla *(Gorilla gorilla)*; orangutan *(Pongo pygmaeus)*; gibbon *(Hylobates* sp.); chimpanzee *(Pan* sp.); baboon *(Papio* sp.); gelada *(Theropithecus gelada)*; rhesus *(Macaca mulatta)*; cynomolgus *(Macaca fascicularis)*; Formosan rock macaque *(Macaca cyclopis)*; Japanese macaque *(Macaca fuscata)*;

---

[1] World Health Organization Collaborating Laboratory on Comparative Medicine: Simian Viruses.

[2] According to Napier and Napier, 1967, the following are the correct species and common names for the *aethiops* group: *C. aethiops* (= grivets); *C. pygerythus* (= vervets); *C. sabaens* (green monkeys).

African green *(Cercopithecus aethiops)*; patas *(Erythrocebus patas)*; talapoin *(Cercopithecus talapoin)*; galago *(Galago* sp.); marmoset *(Callithrix* sp.); squirrel *(Saimiri* sp.); woolly *(Lagothrix* sp.); and owl *(Aotus trivirgatus)*.

Complement-fixation: Because of various problems associated with the serologic testing for antibody to this agent, a brief description of the procedure employed in our laboratory is presented. Specific methods have been described in detail elsewhere [2]. All sera are stored at −20 °C until tested. Prior to testing, the sera are diluted 1 : 10 in veronal buffer and inactivated for 30 minutes at 58 °C. In a number of instances sera were treated with CO_2 and/or heated at 60 °C for 20 minutes.

The microtiter test employed is an adaptation of the macro procedure previously published [2]. Hemolysin (Difco Lab.) is titrated from a 1 : 100 stock solution. Two units are used in the test. Complement (Difco Lab.) is titrated from a freshly prepared 1 : 30 dilution maintained in an ice bath. Two full units are used in the test. Antigen supplied by Dr. R. E. KISSLING (CDC) is an inactivated guinea-pig liver preparation [3]. This antigen was "box" titrated against guinea-pig antisera also provided by Dr. KISSLING. Two units are used in the test. Appropriate controls include a negative guinea-pig serum as well as normal tissue, in this case *uninfected* guinea pig liver.

All tests included positive and control antigens along with a control for anti-complementary (AC) activity by the individual serum. In reading the test, both the reaction with the control antigen as well as the AC reaction of the serum was considered for final evaluation. Accordingly, a serum would be called positive only if it reacts as 3 or 4+ with all the above taken into consideration. Screening was at 1 : 10 and positives along with a representative number of negatives were titrated for endpoint, i.e. 1 : 10 or 1 : 20−1 : 640.

Results

Results obtained in surveying these sera at a 1 : 10 dilution are as follows. Human sera (Table 1) include African natives, laboratory personnel, and military recruits (young adults). None of these sera showed any evidence of antibody.

Table 1. *Number of Human Sera with Antibodies to the Marburg Virus*

Source	No. Pos./ No. Tested
African Natives	0/29
SFRE Lab. Personnel	0/49
Recruits (1967)	0/24
Lab. No. 11 Personnel	0/9

As can be seen, none of 111 "normal" human sera were found to be positive. Included among these were sera from individuals having had extensive exposure to the talapoin *(Cercopithecus talapoin)*. Of the higher apes tested for antibody, positives were found among the gorilla, orangutan, and chimpanzee (Table 2). The gorilla positives were not seroconversions as they had not been tested previously. The orangutans, however, do represent conversions, at least in 4 of the

Table 2. *Number of Gorilla, Orangutan, Gibbon and Chimpanzee Sera with Antibody to the Marburg Virus*

Source	No. Pos./ No. Tested
Gorilla, Lab. No. 1	0/5
1968	3/12
Orangutan, Lab. No. 1, 1963	0/17
1966	0/26
1968	5/19
Gibbons, Lab. No. 4	0/18
Chimpanzee, Lab. No. 1, 1963	0/21
1966	1/35
1966 (Lab Born)	0/27
1968	12/22
SFRE — Pre bleeding	2/15
4th bleeding (Approx. 1 yr.)	6/12
Lab. No. 2, 1964	1/4
1968	14/28
Lab. No. 4, Group 1	2/30
2	3/3
3	0/3

5 positives. Originally the orangutan sera were all negative as indicated. These positive sera all have been retested and titrated for endpoint showing titers of 1 : 10 (3), 1 : 20 and 1 : 40. All the gibbon sera were negative. The chimpanzee sera are interesting inasmuch as they not only represent a group of animals with a high incidence of positives, but also indicate seroconversions from negative to positive in at least two laboratories. There is also some evidence of antibody present in chimpanzees bled at time of capture (SFRE prebleeding 2/15). These animals were bled in Africa but the exact time with relation to capture is not clear.

The large number of chimpanzee positives at laboratories no.'s 1 and 2 is not explainable at this time. Both facilities have many other simians and contact among the various groups is frequent and commonplace. Positive titers of the Lab. No. 2 animals ranged from 10 to 80. Interestingly, and not known at the time of testing, was the fact that there were two sets of duplicates among these sera. In both instances these sera gave essentially the same titers (1 : 20—1 : 10 and 1 : 80—1 : 80).

The baboon data (Table 3) are important because these are the only animals in the entire study on which blood samples were obtained, in certain circumstances, immediately following capture. There is, therefore, strong evidence to suggest, at least for baboons, that this infection occurs in Africa.

In a number of instances, baboon troops sampled immediately upon capture were found without antibody to this antigen. Seroconversions were observed among two baby baboons, born into the SFRE colony which became positive

Table 3. *Number of Baboon Sera with Antibodies to the Marburg Virus*

Source	No. Pos./ No. Tested
African — SFRE[1]	0/13
Domestic — SFRE[2]	0/18
Kenya — 1963[3]	0/22
Kenya — 1964[3]	3/7
Kenya — 1966[3]	8/88
Kenya — 1968[3]	9/92
Newborns — SFRE	0/17
Approx. 6 months	2/20
Baseline — SFRE	1/21
Approx. 1 year	0/6
Baseline — SFRE	1/10
Approx. 6 months	4/9
Lab. No. 5	0/18
Lab. No. 4 (Geladas)	0/10

[1] Animals born in Africa but maintained at SFRE for several years prior to testing.
[2] An "inbred" colony of baboons born in San Antonio.
[3] Represents field trips to Africa and bleeding of animals immediately upon capture.

after 6 months (approximately). Similar conversion was noted in one baseline group of animals with 4 of 9 animals becoming positive after approximately 6 months in captivity. It is possible, however, that this latter group was infected at the time of capture (and first sampling) but the serologic test was negative. These animals then were detected as positive upon testing their sera 6 months later. It may be noted that the two colonies long established at SFRE (one group consisting of only baboons born in captivity, the other at least 5 years in residence) were both negative.

Baboon sera investigated or evaluated in cooperation with Dr. MALHERBE will be discussed later.

Data on other Old World monkey sera are presented in Table 4. The rhesus group representing 166 animals at 9 different facilities indicated an incidence of 7.8% positives. One group at Lab. No. 3 showed some seroconversion from negative to positive in 3 animals over a two year period. Of the Old World Asian monkeys, these animals demonstrated the lowest incidence of antibody.

As can be seen by examining 3 other groups of *Macaca*, there is evidence of antibody in all the groups tested. The two small groups of cynomolgus and Formosan rock macaque monkeys were somewhat similar. In view, however, of our previous assertion relative to finding antibody predominantly in African animals, the findings with the Japanese macaques were somewhat surprising. It had been rationalized that the occasional positive in the other Asian simians was possibly related to contact with African animals. There was some justification for this opinion as a number of servoconversions had been found among animals maintained in captivity.

Table 4. *Number of Old World Monkey (Asian) Sera with Antibody to the Marburg Virus*

Source	No. Pos./ No. Tested
Rhesus SFRE	0/18
Lab. No. 3 — 1966	0/15
1968	3/22
Lab. No. 4	3/20
Lab. No. 10	1/23
Lab. No. 9	0/16
Lab. No. 19 (Group 1)	1/11
(Group 2)	0/9
Lab. No. 22	1/20
Lab. No. 23	4/8
Lab. No. 27	0/4
Cynomolgus — Lab. No. 5 (Group 1)	2/16
(Group 2)	1/12
Lab. No. 19 (Group 1)	3/12
(Group 2)	3/13
Formosan Rock macaque — Lab. No. 16	4/25
Japanese macaque — Lab. No. 17	15/23

Examination of these Japanese macaque sera for antibody to other agents indicated that antibody to an uncommonly large number of antigens are detected by CF and HI procedures. At this time it is difficult to evaluate these findings. It would appear that these animals do have a rather high incidence of "antibody" as a result of 1. numerous contacts with various agents including the Marburg virus, 2. a nonspecific inhibitor in the sera of this species of *Macaca* resulting in a high incidence of "seropositives", or 3. contact with other simians causing infection with this agent. Further studies are obviously required to evaluate this finding. It might be pointed out at this time that similar findings were obtained when a group (not previously tested) of immature Japanese macaque (9/18—50%) were similarly tested. Repeats on these sera, along with titrations for endpoint, confirmed these original positive findings. Dr. T. TANAKA, who kindly provided these sera, indicates that many of these animals have had contact prior to capture with a great many visitors.

Testing Old World African simians for antibody (Table 5) continued to indicate large numbers of serologic positives with the talapoin still showing the greatest incidence. Variation in antibody incidence among the different vervet colonies was quite apparent. Seroconversions among the SFRE animals may be noted.

The talapoin groups are two separate test groups and do not represent seroconversions. A small group of galago (Prosimian) sera was included in the test with negative findings.

Table 5. *Number of Old World Monkey (African) Sera with Antibody to the Marburg Virus*

Source	No. Pos./ No. Tested
African green — SFRE (Group 1)	2/19
(Group 2)	4/19
Lab. No. 5	2/16
Lab. No. 11 (Group 1)	5/21
(Group 2)	0/6
Lab. No. 1	1/2
Lab. No. 27	0/3
Lab. No. 25 — Plastic	15/40
Rubber	17/25
Patas Lab. No. 5	2/23
Lab. No. 4	4/15
Talapoin Lab. No. 11 (Group 1)	9/14
(Group 2)	15/20
Galago (Prosimian) Lab. No. 11	0/24

Approximately 76 New World simian sera were tested representing major species employed in various laboratories (Table 6). It would appear that the Marburg agent has not been in contact with this group with the possible exception of one marmoset known to be in contact with other primates and animals.

Table 6. *Number of New World Monkey Sera with Antibody to the Marburg Virus*

Source	No. Pos./ No. Tested
Marmosets[1] Lab. No. 8	0/16
Lab. No. 14 (Group 1)	0/3
(Group 2)	1/13
Lab. No. 18	0/6
Squirrel Lab. No. 18	0/2
Lab. No. 22	0/10
Woolly Lab. No. 18	0/6
Owl Lab. No. 18	0/8
Mixture of animals[2] Lab. No. 21	0/12

[1] Different genus and species.
[2] Represents marmosets, squirrel, woolly and owl monkeys not included in the above.

Table 7 summarizes the findings on the various simians tested. The animals are listed geographically as well as by genus and species.

At this time a number of questions relative to the specificity of this reaction arose. Sera were obtained through the courtesy of Dr. SIMPSON from a number of hams-

Table 7. *Incidence of Positive Sera (Marburg Virus) in Sera according to Genus and Species and Geographic Distribution*

	No. Pos./ No. Tested	%
Old World		
African		
Gorilla *(Gorilla gorilla)*	3/17	1.8
Chimpanzee *(Pan sp.)*	41/200	22.4
Baboon *(Papio sp.)*	32/370	8.7
African green *(Cercopithecus aethiops)*	34/126	27.0
Patas *(Erythrocebus patas)*	6/38	15.8
Talapoin *(Cercopithecus talapoin)*	24/34	70.6
Asian		
Orangutan *(Pongo pygmaeus)*	5/62	8.1
Gibbons *(Hylobates* sp.)	0/18	—0—
Rhesus *(Macaca mulatta)*	13/166	7.8
Cynomolgus *(M. irus)*	8/53	15.1
Formosan Rock Macaque *(M. cyclopis)*	4/25	16.0
Japanese Macaque *(M. fuscata)*	15/23	65.2
New World		
Marmosets *(Callithrix, Saguinus, Leontideus, Cebuella)*		
Squirrel *(Saimiri sciureus)*	1/76	1.3
Woolly *(Lagothrix lagothrica)*		
Owl *(Aotus trivirgatus)*		

ters and guinea pigs. These results (Table 8) are presented with titers ranging from 1 : 20 — 1 : 320. Nothing is known about these sera other than that they were marked "immune sera".

Table 8. *Antibody to Marburg Virus in Hamster and Guinea pig Sera*[1]

		Serum Dilution					
Source	Animal No.	20	40	80	160	320	640
Hamster	(1) — 4619	4	4	4	2	1	1
	(2) — 4620	4	4	4	4	1	1
	(3) — 4621	4	4	4	2	1	1
	(4) — 4622	4	4	4	4	2	1
	(5) — 4623	4	4	4	4	3	1
	(6) — 4624	4	4	4	4	3	1
	(7) — 4625	4	4	3	3	1	1
	(8) — 4626	4	4	4	4	2	2
	(9) — 4627	4	4	3	3	2	1
	(10) — 4628	4	4	4	2	1	1
Guinea pig	(1) — 4629	4	4	3	2	1	1
	(3) — 4630	4	4	3	3	1	1
	(5) — 4631	4	4	4	4	2	1
	(6) — 4632	4	4	3	2	1	1
	(7) — 4633	3	2	1	1	1	1
	(9) — 4634	4	3	3	3	1	1

[1] Supplied through the courtesy of Dr. SIMPSON.

Continued concern regarding these findings prompted additional studies. Sera kindly supplied by Dr. M. SIEGERT on 3 patients recovered from the outbreak were subject to the sera test system (Table 9). On the basis of these results as well as

Table 9. *CF Results on 3 Human Patient Sera with Marburg Virus*

Patient	20	40	80	160	320	640
No. 1	1	1	1	1	1	1
No. 2	4	3	2	2	2	2
No. 3	4	3	2	2	1	1

two additional tests, sera no.'s 2 and 3 were considered positive. These results are of interest inasmuch as they were obtained May 5, 1969, May 5, 1969, and April 23, 1969, respectively. Thus, approximately two years following occurrence of the disease, CF antibody is still present in 2 of 3 patients.

Continuing with our attempt to understand the problem and evaluate the results, a cooperative study was instituted with Dr. MALHERBE. Results on a number of these sera are presented in Table 10. These sera from vervets were sent to us as split samples because of a question relative to a possible nonspecific effect due to the container. As can be seen, from the comparative tests, some differences were obtained with sera in plastic and rubber stoppered tubes. Also, differences were obtained between the two laboratories on the same specimen. Incidently, these data are only related to tests performed with the same antigen supplied by Dr. KISSLING. Dr. MALHERBE has prepared his own antigens and has discussed his results with that antigen on these sera.

It is obvious that a large number of sera were either nonspecific, anticomplementary, or both, when tested in the two laboratories. Also little agreement was obtained in our laboratory by testing the sera stoppered differently. However, a number of sera were found to be positive when tested in Johannesburg and San Antonio. As a result, possible agreement with regard to positives was obtained in 12 of 50 samples. Good agreement was noted for the negatives in both laboratories. The essential problem centers around sera that demonstrated some nonspecific or anticomplementary activity (or both).

Discussion

What does all this mean? In our opinion while there are many problems to be resolved regarding interpretation of these data from many parameters, this CF procedure does detect specific antibody. Obviously, while a crude antigen such as the one prepared by Dr. KISSLING, will contain a number of unwanted antigens, specific antigen undoubtedly is also present. This conclusion is derived from the following observations as a result of this study:

1. Sufficient positive sera have now been tested to justify the opinion that antibody is detected.

2. Comparative testings, while not in the most desirable range of agreement, does indicate some conformity.

Table 10. *CF Results on Vervet Sera as Tested in Johannesburg and San Antonio*

	Jo'burg		San Antonio Plastic			Rubber		
Animal No.	+ Ant.	— Ant.	+ Ant.	— Ant.	AC	+ Ant.	— Ant.	AC
1 — 2371	2	±	160	80	—	80	20	—
2 — 2372	1	±	—	—	—	—	—	—
3 — 2373	2	±	—	—	—	—	—	—
4 — 2374	2	±	20	—	—	40	40	—
5 — 2375	4	±	40	40	20	40	40	—
6 — 2376	--	—	40	40	40	—	—	—
7 — 2377	1	—	—	—	—	—	—	—
8 — 2378	—	—	—	—	—	10(?)	—	—
9 — 2379	±	±	—	—	—	—	—	—
10 — 2380	1	+	—	—	—	—	—	—
11 — 2381	2	±	—	—	—	40	40	40
12 — 2382	2	±	80	80	80	20	—	—
13 — 2383	2	±	—	—	—	—	—	—
14 — 2384	4	4	—	—	—	—	—	—
15 — 2385	4	±	40	—	—	40	20	—
16 — 2386	4	2	—	—	—	20	20	—
17 — 2387	2	1	—	—	—	40	20	—
18 — 2388	4	2	20	20	—	20	20	—
19 — 2389	4	±	—	—	—	40	—	—
20 — 2390	4	3	20	20	—	20	20	—
21 — 2391	4	1	10	—	—	40	—	—
22 — 2392	4	2	—(?)	—	—	40	—	—
23 — 2393	3	1	—	—	—	—	—	—
24 — 2394	4	±	20	20	10	40	—	--
25 — 2395	4	1	40	40	40	—(?)	—	—
26 — 2396	4	1	—	—	—	20	20	—
27 — 2397	4	±	40	20	—	40	—	—
28 — 2398	4	2	10(?)	—	—	160	40	—
29 — 2399	4	1	—	—	—	4	3	—
30 — 2400	4	±	10(?)	—	—	10(?)	—	—
31 — 2401	4	±	20	—	—	20	20	—
32 — 2402	3	±	—	—	—	—	—	—
33 — 2403	4	±	20	20	—	—	—	—
34 — 2404	4	—	20	20	20	40	—	—
35 — 2405	4	±	?	?	?	?	?	?
36 — 2406	3	—	—	—	—	—	—	—
37 — 2407	4	1	20	—	—	10(?)	—	—
38 — 2408	4	1	40	40	40	20	20	—
39 — 2409	3	1	—	—	—	—	—	—
40 — 2410	3	—	20	20	20	40	40	20
41 — 2411	4	±	—	—	—	20	—	—
42 — 2412	4	—	10	10	10	40	—	—
43 — 2413	2	—	—	—	—	—	—	—
44 — 2414	4	±	10	—	—	10	10	—
45 — 2415	4	±	10	—	—	40	40	—
46 — 2416	4	—	X	X	X	10	10	—
47 — 2417	4	—	X	X	X	160	40	—
48 — 2418	4	—	—	—	—	10	10	—
49 — 2419	4	3	X	X	X	20	—	—
50 — 2420	4	3	80	80	80	160	160	80

3. Distribution of antibody, while still not finalized does suggest a definite pattern with a primary locus in Old World nonhuman primates and strong evidence for lack of antibody in New World monkeys.

Other attempts to compare results were also included in this study. Approximately 31 simian sera previously tested at SFRE were sent to Dr. SIMPSON for testing in his laboratory. Included in this group were vervet and talapoin sera found to be positive as well as negative in our hands. Dr. SIMPSON reported all but 4 sera were AC although many attempts to remove this AC activity were employed. The 4 that were satisfactory were negative in his hands. These 4 sera were also negative in our tests.

Six sera (2 talapoin, 2 Japanese macaque and 2 baboon) were sent to Dr. MALHERBE. In this small group, better correlation apparently was obtained. Of the two talapoin sera, both with a titer of 1 : 40 in our hands, one was positive at 1 : 5 and 1 : 10 with the Kissling antigen in Johannesburg. The two Japanese macaque sera having titers of greater than 1 : 320 in our hands were both positive at 1 : 5 and 1 : 10. Our two negative baboon sera were both NS/AC in Dr. Malherbe's lab. It should be emphasized that all six sera were either negative or NS/AC when tested with Dr. Malberbe's monkey antigen with the possible exception of one Japanese macaque serum. Repeat testing of this "positive" serum could not confirm the original result.

These results probably raise more questions than answers. Further testing using other antigen preparations would help resolve the discrepancies reported between Dr. Malherbe's laboratory and ours. Perhaps a tissue culture antigen preparation, as reported by Dr. SLENCZKA at this meeting, would be helpful. Furthermore, the enhanced nonspecific and anticomplementary activity noted as a result of shipping these test sera to England and from South Africa to San Antonio requires investigation.

Experimental infection apparently results in 100% fatality. If this situation occurs in nature, obviously there could not be any serologic positives. However, if the experimental route of inoculation differs from that occurring in nature, it is possible that 100% fatality will not develop.

The suggestion has been made that the antigen employed here detects an antibody other than that due to the Marburg virus. This possibility is conceivable but additional studies, as mentioned above, are required for further elucidation. Meanwhile, in the routine performance of various serologic procedures, a necessary competence is developed by the individuals performing the test. If, in the minds of these individuals, after all factors and vagueness associated with the serologic procedures employed are taken into consideration, the test represents a positive value then it must be reported as such. Lack of reliance on the tests employed would strongly suggest their discontinuation.

Acknowledgements

This study was funded, in part, by U. S. P. H. S. grants No. 1-PO6-FROO361 and 5-PO6-FROO278 and W. H. O. grant No. Z2/181/27 to Southwest Foundation for Research and Education.

We are appreciative to the following facilities for supplying the sera used for this study: Animal Farm Division, Fort Detrick, Maryland; Animal Care Facility,

University of California, San Francisco, California; Auburn University, Auburn, Alabama; 6571st Aeromedical Research Laboratory, Holloman A. F. B., New Mexico; U. S. A. F. School of Aerospace Medicine, Brooks A. F. B., Texas; Naval Aerospace Medical Institute, Naval Aerospace Medical Center, Pensacola, Florida; Dental Science Institute, The University of Texas Dental Branch, Houston, Texas; Epidemiology and Research Analysis Section, Foreign Quarantine Program, National Communicable Disease Center, Miami, Florida; Delta Regional Primate Research Center, Covington, Louisiana; Institute Merieux, Lyon, France; Institute for Comparative Biology, Zoological Society of San Diego, San Diego, California; Department of Medicine, Division of Comparative Medicine, University of Florida College of Medicine, Gainesville, Florida; Japan Monkey Center, Institute of Primatology, Aichi, Japan; New England Regional Primate Research Center, Southborough, Mass.; Presbyterian-St. Luke's Hospital, Chicago, Illinois; National Taiwan University College of Medicine, Taipei, Taiwan, China; Yerkes Regional Primate Research Center, Atlanta, Georgia; Research Foundation, Johannesburg, South Africa; Hygiene-Institute der Universität, Marburg, Germany; Microbiological Research Establishment, Porton Down, Salisbury, Welts, England.

We also wish to acknowledge the able technical assistance of Mrs. J. J. RATNER, BETTYE TUNMER, and DONNA THOMAS.

References

1. KALTER, S. S., RATNER, J. J., HEBERLING, R. L.: Antibodies in primates to the Marburg virus. Proc. Soc. Explt. Biol. and Med. **130,** 10 — 12 (1969).
2. KALTER, S. S.: Procedures for routine laboratory diagnosis of virus and rickettsial diseases. Minneapolis, Minnesota: Burgess Publ. Co. (1963).
3. KISSLING, R. E., ROBINSON, R. Q., MURPHY, F. A., WHITFIELD, S. G.: Agent of disease contracted from green monkeys. Science **160,** 880 — 890 (1968).

Studies on the Marburg Virus

H. MALHERBE and M. STRICKLAND-CHOLMLEY

With 2 Figures

The events and circumstances of the outbreak of Marburg virus disease in Germany and Yugoslavia during August and September, 1967, have been fully documented by a number of authors in the Deutsche Medizinische Wochenschrift, March 26, 1968, and in other publications [1, 2, 3].

We briefly reported our preliminary findings in 1968 [4], because we felt that it may not have been sufficiently appreciated that the virus could multiply undetected in monkey kidney tissue cultures unless the cells were stained to reveal the virus-induced cytoplasmic inclusions. The present communication summarizes our investigations to date, with the exception of the examination of South African baboons and vervet monkeys for natural infection with Marburg virus, which is the subject of a separate paper [5]. Our objectives have been clearly defined on a practical basis:

1. To establish the identity of the virus isolates made in our laboratory,
2. To assess techniques for the rapid and reliable detection of the virus,
3. To provide additional information of value in the handling of this agent.

Virus Isolations

Our first isolation was made in tissue cultures prepared from the kidney of the vervet monkey Cercopithecus (aethiops) pygerythrus, which had been inoculated with serum from the patient KLIEBE, sent in 1967 by Professor R. SIEGERT for serological investigation. The virus was re-isolated twice from this serum.

Through the courtesy of Professor SIEGERT and Dr. GORDON SMITH, further samples of blood and tissues were obtained; and we recovered (and re-isolated) similar viruses from the bloods of patients HARTZ and FLAK, and from the liver of a rhesus monkey infected at Porton.

Laboratory Facilities

Although we were accustomed to handling a variety of simian viruses in our laboratory, special facilities for dealing with this particularly dangerous agent had to be improvised. A steel hood for tissue culture and complement fixation work was made, the air being drawn in through open front ports and passed out through filters irradiated by ultraviolet lamps. The tissue culture work involved 2800 cultures, of which more than 1000 contained multiplying virus; and over 2000 stained coverslip cultures were examined.

For studies in animals, we emptied an experimental monkey wing and used rooms at the furthest end. Sealed plastic enclosures for cages were improvised: 2 monkeys in separate cages, and up to 8 small animal cages, could be accommodated, accessible by way of tape-sealed plastic doors. Precautions were taken to prevent cross-infection between monkeys and small animals, and air was constantly under negative pressure in the cages. Evacuated air was drawn first over heated coils, then along three 36-inch ultraviolet lamps in series, and finally into a box irradiated by four 36-inch ultraviolet lamps before being discharged into the air-conditioning exhaust duct.

Tissue Cultures

Except for the preparation of bulk virus stocks, roller tube cultures containing free coverslips were maintained at 36—37 °C. Various media appropriate for the different tissues tested were used, with foetal calf serum added. Most of the

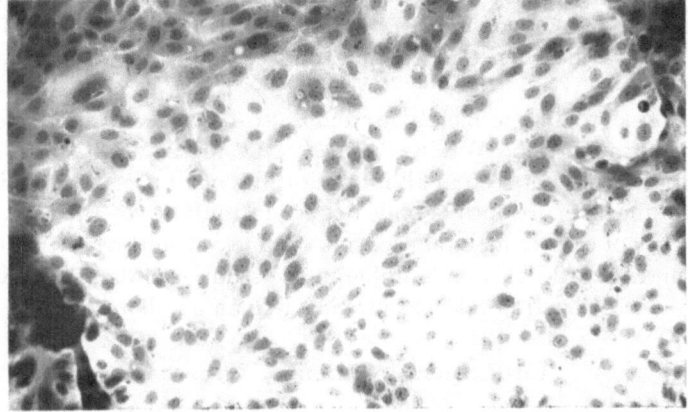

Fig. 1. Vervet monkey kidney tissue culture 14 days after infection with Marburg virus. Magnified 200 ×

coverslips were fixed with Bouin's solution followed by 70% ethyl alcohol, and were stained with haematoxylin and eosin (or eosin and phloxine).

Our first isolation had shown that while an infected sheet of monkey kidney cells might appear intact (Fig. 1), staining of the cells and high-power microscopic examination would reveal eosinophilic cytoplasmic inclusions (Fig. 2).

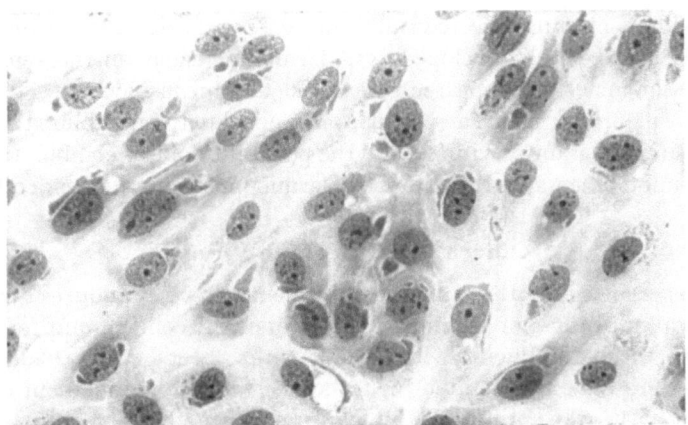

Fig. 2. Same culture as shown in Fig. 1, magnification 480 ×

The earliest evidence of infection was found on the second day after inoculation, but this was exceptional; and usually inclusion formation was conspicuous only after the seventh day. When very small amounts of virus were present, scanty foci might be seen only after 2 weeks, but a culture held for 20 days usually sufficed to reveal the virus adequately. However, in order to avoid the possibility of cross-contamination when replacing medium in titration tests, we titrated virus in tubes which remained unopened for 10—12 days, after which period the test was terminated and the coverslips stained. In more critical experiments, such as the detection of small amounts of virus after exposure to heat, cultures were maintained for 3 weeks with a single medium change on the tenth day.

In our hands, primary vervet kidney cultures proved to be hardy and suitable for practical purposes. Secondary vervet kidney cells were also successfully used, but these increase the chance of complicating experiments with contaminating simian viruses.

Eosinophilic cytoplasmic inclusions could also be detected in infected continuous line cultures of HeLa, WI-38, BHK-21, LLC-MK2, VERO, and AGM-AH cells. Inclusions were not seen in all infected tubes of LLC-MK2 cells, indicating that this line may be susceptible only to certain mutants of the virus. In our experience it is not easy to maintain continuous line cell cultures without frequent changes of medium, which is undesirable in the study of a dangerous virus. No inclusions could be found in infected primary human amnion cells.

Co-Existence with other Viruses

The possible contamination of poliovirus vaccine with Marburg virus led us to enquire whether poliovirus would multiply in Marburg virus-infected cells. Primary vervet kidney cultures were challenged with 100 $TCID_{50}$ of the oral vaccine strain LSC-2AB Type 1 poliovirus 17 days after they had been infected with a high dose of Marburg virus. As compared with cultures receiving poliovirus only in parallel, the Marburg virus-infected cultures showed a 24-hour delay in the development of poliovirus cytopathic effect, but staining of the cultures demonstrated that Marburg virus inclusions could be found in poliovirus infected cells.

Vervet kidney cultures infected simultaneously with foamy virus and Marburg virus usually showed well-developed cytoplasmic inclusions in the syncytia, but occasionally the inclusions appeared thread-like and inconspicuous. Foamy virus combined with Marburg virus can produce an effect closely resembling that of some human and simian paramyxoviruses, and the possibility of this combination should be borne in mind when syncytia and cytoplasmic inclusions occur together.

Other Properties of the Virus

In our experience only the human parainfluenza Type 1 produces cytoplasmic inclusions similar to those of Marburg virus without the concomitant formation of syncytia, but the parainfluenza virus destroys the sheet of vervet kidney cells. Paramyxoviruses have been stated [6] to be resistant to actinomycin: we found that Marburg virus was not inhibited in cultures treated with medium containing 0.1 microgram of actinomycin D per ml. However, we failed to obtain haemadsorption or haemagglutination by Marburg virus using erythrocytes from 8 species of animals, and the disease produced by the virus also sets it apart from known paramyxoviruses.

One part of ether or chloroform added to 3 parts of cell-free Marburg virus suspension appeared to eliminate all infectivity after overnight exposure at 4 °C. Exposing the virus to sodium deoxycholate solution at a final concentration of 0.2% for one hour at 37 °C reduced the infectious titre by more than 3 logs. Infectivity was not abolished by aluminium chloride at a final 0.2 millimolar concentration in the medium, as used by WALLIS and MELNICK [7] for the suppression of certain adventitious agents in monkey kidney cultures.

Exposure to an ultraviolet sterilamp for 30 seconds at a distance of 10 cm appeared to destroy infectivity. Heating cell-free suspensions of Marburg virus at 56 °C for 30 minutes, or at 60 °C for 10 minutes left a small amount of heat-resistant virus; but heating at 60 °C for 20 minutes eliminated infectivity in the tests we carried out. The virus was not protected by molar magnesium chloride against the effects of heating at 50 °C for one hour.

Growth of the virus was not stimulated by lowering or raising the temperature of incubation; nor did cortisone acetate at a concentration of 20 micrograms per ml in the medium enhance viral growth.

We were able to pass the virus through a Gradocol membrane of 820 millimicrons a. p. d., but not through one of 300 millimicrons a. p. d., under 2—5 lbs positive pressure. Electron microscopy of infected tissues prepared by us was done by Dr. LECATSAS at the Onderstepoort Veterinary Research Institute, but the large rod-shaped bodies described by other workers were not observed.

Virus stocks were prepared in primary vervet monkey kidney tissue and stored at −70 °C. This material was checked for leptospira and mycoplasma, with negative results. The inclusions in monkey kidney cells were Feulgen negative.

Complement Fixation Test Antigens

Several attempts were made to produce complement fixing antigens by culturing Marburg virus in vervet kidney monolayer cultures and in flask cultures of minced monkey liver or spleen. Despite a number of variations in culture technique, a useful antigen was not obtained. However, in view of the advantages of tissue culture for the production of relatively clean preparations of virus, and for permitting an uninfected portion of the same tissue to serve as negative control antigen, it is felt that attempts to exploit the technique should not be abandoned.

Animal Inoculation

1. Suckling Mice

Seventy-two 48-hour-old white mice were inoculated; half by the combined subcutaneous and intracerebral route, and half by the intraperitoneal route. No overt illness resulted. Fifty-seven of the mice were sacrificed at intervals from 3 to 40 days after inoculation, for histological examination. No lesions attributable to Marburg virus were seen except in the brain of one mouse taken on the 8th day after intracerebral inoculation. A necrotic lesion was found, and eosinophilic cytoplasmic inclusions resembling those of Marburg virus were present in a number of neurones adjacent to the lesion. After day 14, no lesions could be detected in mice inoculated intracerebrally. In both inoculated and control uninoculated mice we observed round cytoplasmic inclusions in neurones, particularly near the cerebellar peduncle, which closely resembled the inclusions previously described by one of us [8] as occurring in the cerebral neurones of vervet monkeys.

2. Guinea pigs

First kidney culture passage virus was inoculated intracerebrally into a guinea pig with no apparent effect. Second monkey kidney passage virus inoculated intraperitoneally into 2 guinea pigs produced a rise in temperature above 104 °F in one animal only, on the 8th day. Third monkey kidney passage virus inoculated intraperitoneally into 3 guinea pigs produced pyrexia in all three by the 7th day. These animals were sacrificed and showed focal liver lesions with characteristic eosinophilic degeneration of parenchymal cells.

In order to reduce the possibility of involving unwanted micro-organisms through repeated passage in animals, we decided against preparing complement fixing antigens in guinea pigs after guinea pig liver of the third animal passage proved to be too weak as an antigen in spite of histological evidence of Marburg infection.

An attempt was made, however, to produce antiserum in guinea pigs according to the method recommended by Kissling and his colleagues [9], using 0.7 ml of 1 : 2000 beta-propiolactone inactivated guinea pig liver suspension followed 13 days later by 0.1 ml of infective liver suspension. Of the 6 animals used, 4 died between the 7th and 9th days following the live challenge. The fifth guinea pig was moribund on the 9th day and was sacrificed. Marburg virus was recovered from the spleen and liver, and the characteristic focal eosinophilic degeneration of liver cells was found. Similar lesions were not found in uninfected control guinea pigs, nor have they previously been seen by us in guinea pig livers.

Six guinea pigs inoculated intraperitoneally with 2nd monkey kidney passage virus showed pyrexia from the 4th to the 8th day. One died on the 9th day, and the remaining 5 were bled from the heart and sacrificed. All of these 5 showed the typical liver lesions, and Marburg virus was recovered from the blood of 2 of the animals.

3. Monkeys

Two monkeys weighing just over 2 kilos were each given 1 ml of 7th monkey kidney passage virus, comprising a dose of 10^4 $TCID_{50}$, administered into the nose and mouth. One of these vervet monkeys died on the 11th day: the spleen was extensively necrosed; two-thirds of the liver cells showed eosinophilic degeneration; and in the adrenals were seen scattered eosinophilic necrotic cells and patches of haemorrhage. Marburg virus was recovered from the liver, spleen, and kidney of this animal in tissue cultures, but not from the throat swab taken on the day of death. The second monkey remained well and was sacrificed on the 14th day: viral changes were not noted in the liver or spleen, and virus was not recovered from them.

In a second experiment 2 vervet monkeys each weighing 5.75 kilos were inoculated with the intention of producing hyperimmune serum. The virus given had been through one VERO cell passage followed by 3 vervet kidney passages, and the fluid contained approximately 10^5 $TCID_{50}$ per ml. Two intramuscular inoculations, each consisting of 2 ml of virus fluid inactivated with 1 : 2000 beta-propiolactone, emulsified in 2 ml of mineral oil, were given with a 12-day interval. After a further 12 days, each animal received 2 ml of untreated virus fluid intramuscularly. Of these two monkeys, that numbered D 788 became progressively ill from the second day after live challenge, and it died on the 9th day. Focal lesions were noted in the brain and adrenals; the spleen was virtually entirely necrosed; and more than half of the liver cells showed degeneration. Phloxinophilic inclusions were demonstrated in the liver cells with Lendrum's stain. Marburg virus was recovered in tissue cultures from the spleen and urine, but the liver was too toxic to tissue cultures for immediate virus recovery. Liver tissue from this monkey D 788 was stored at -70 °C, and a portion of it was subsequently used, after fluorocarbon treatment, as antigen for complement fixation tests described by us in another paper. The antigen thus treated had an infectious titre of $10^{3.0}$ $TCID_{50}$ per ml.

The second monkey in this test, numbered D 787, appeared ill from the second day following live challenge, but it survived and was exsanguinated on the 16th day. This serum was used as the immune monkey serum in our complement fixation tests.

Discussion

Of the four strains of Marburg virus isolated by us, only those recovered from the bloods of patients KLIEBE and HARTZ were used in the above experiments. In our hands, primary vervet monkey kidney tissue cultures proved more satisfactory for virus detection than did continuous cell line cultures or the mice and guinea pigs used.

Since Marburg virus is capable of multiplying in monkey kidney cells without producing a gross cytopathic effect, the staining of cultures should be employed by all who use primary simian tissues, and it should be mandatory in the safety testing of vaccine produced in monkey tissue.

Acknowledgements

We are grateful to Professor R. SIEGERT and Dr. F. LEHMANN-GRUBE of Marburg University, and to Dr. GORDON SMITH and his colleagues at Porton, for so generously supplying us with materials for virus isolation.

We wish to thank Dr. J. H. S. GEAR, Director of the Poliomyelitis Research Foundation, for allowing these studies to proceed; Mrs. Z. KIRSCH for preparing the histological sections; and Mr. M. ULRICH for the two photographs.

References

1. World Health Organization: Wkly Epidem. Rec. **42,** 479 (1967).
2. GORDON SMITH, C. E., SIMPSON, D. I. H., BOWEN, E. T. W., ZLOTNIK, I.: Lancet **2,** 1119 (1967).
3. SAENZ, A. C.: Primates in Medicine, **3,** 129 (1969). New York: S. Karger.
4. MALHERBE, H., STRICKLAND-CHOLMLEY, M.: Lancet **1,** 1434 (**1968**).
5. STRICKLAND-CHOLMLEY, M., MALHERBE, H.: To be published.
6. JAWETZ, E., MELNICK, J. L., ADELBERG, E. A.: Review of Medical Microbiology, 7th ed., Oxford: Blackwell. 1966, p. 289.

7. Wallis, C., Melnick, J. L.: Texas Reports on Biology and Medicine. **20,** 465 (1962).
8. Malherbe, H.: Nature, **209,** 832 (1966).
9. Kissling, R. E., Robinson, R. Q., Murphy, F. A., Whitfield, S. G.: Science, **160,** 888 (1968).

Summary

Strains of virus were isolated in tissue cultures from the bloods of 3 patients with Marburg virus disease, and one strain was recovered from the liver of a rhesus monkey inoculated with the virus in England. The different strains all produced eosinophilic cytoplasmic inclusions in primary vervet monkey kidney tissue cultures without destroying the cell sheet, and presence of the virus was only detected when the cultures were stained. Similar inclusions were produced in a variety of continuous cell line cultures. The inclusions were Feulgen-negative.

Properties of the virus included inactivation by ether and chloroform, and by sodium deoxycholate. Heat at 60 °C for 20 minutes destroyed infectivity, but a small heat-resistant fraction remained after exposure to heat at 56 °C for 30 minutes or 60 °C for 10 minutes. Ultraviolet irradiation removed infectivity. The virus was not sensitive to aluminium chloride.

The inoculation of 48-hour-old mice did not result in overt disease, but a lesion suggesting limited viral multiplication was noted in the brain of one mouse inoculated intracerebrally. In guinea-pigs the virus caused pyrexia and characteristic eosinophilic degeneration of the liver cells. One of two vervet monkeys inoculated by the naso-pharyngeal route contracted the disease. In monkeys the virus produced extensive necrosis of the spleen, eosinophilic degeneration of liver cells, and focal lesions in the brain and adrenals. Virus could be recovered from monkey liver, spleen and urine. Inoculation of guinea-pigs and monkeys with 1 : 2000 beta-propiolactone inactivated virus did not protect against subsequent challenge with live virus.

Attempts to produce complement fixing antigens in tissue cultures were not successful, but fluorocarbon treated monkey spleen and liver provided potent antigens. The liver antigen thus treated had an infectious titre of 10^3 TCID$_{50}$ per ml.

Examination of South African Primates for the Presence of Marburg Virus

M. STRICKLAND-CHOLMLEY and H. MALHERBE

With 1 Figure

Studies of antibody to Marburg virus in primates have been reported by SIMPSON et al. [1] and by KALTER et al. [2]. Our investigation was undertaken to assess whether antibody to Marburg virus could be detected in the primates of South Africa, and whether there is other evidence of the occurrence of the virus in this country.

Materials and Methods

Since 1967 sera have been collected from the chacma baboon (Papio ursinus) and the vervet monkey (Cercopithecus (aethiops) pygerythrus) [3]. Fig. 1 indicates approximately 30 areas where baboon and monkey sera were collected, comprising much of the Cape Province and several widely separated areas elsewhere in South Africa.

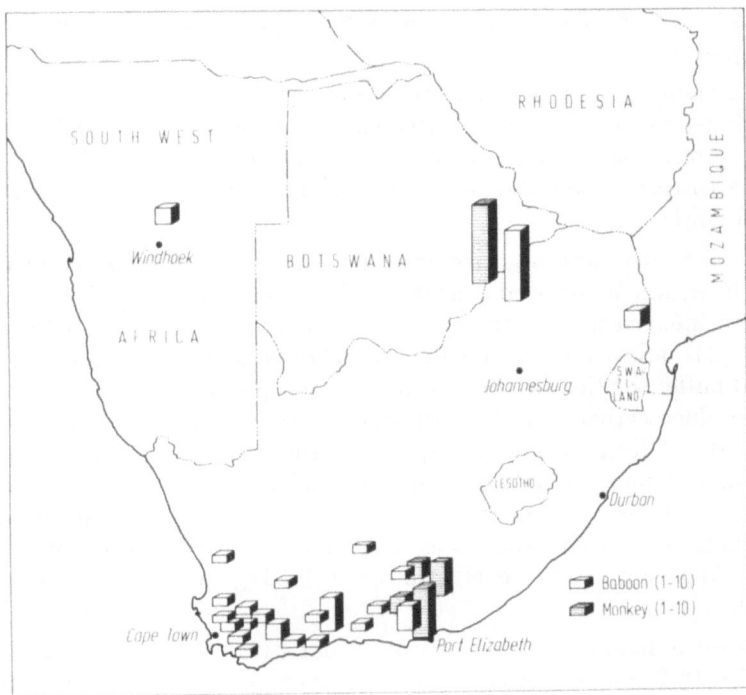

Fig. 1

Sera were stored at −20 °C until used. Those selected for virus isolation were inoculated undiluted into vervet monkey kidney monolayer tissue cultures, and those used for complement fixation tests were diluted 1 : 10 in veronal buffer solution pH 7.4 before inactivation at 56 °C for 30 minutes.

Throat swab suspensions were made in an isotonic buffer solution containing antibiotics, and centrifuged at 2500 r.p.m. for 30 minutes in an anglehead centrifuge; the supernatant fluid then being inoculated into tissue cultures.

The cultures used were of primary vervet monkey kidney monolayers in roller tubes containing free coverslips, and were maintained in Hanks' balanced salt solution—lactalbumen hydrolysate medium for up to 21 days at 37 °C. Coverslip cultures were fixed with Bouin's solution followed by 70% ethyl alcohol, stained with haematoxylin and eosin (or eosin and phloxine), and examined for the presence of the eosinophilic cytoplasmic inclusions characteristically produced by the Marburg virus.

The immune monkey serum was that of monkey No. D 787 described by Malherbe and Strickland-Cholmley [4]. This serum was initially slightly anticomplementary and was treated by diluting it 1 : 9 with distilled water, adding a pellet of carbon dioxide ice approximately 1 cu. cm in size, and centrifuging at 2000 r.p.m. for 10 minutes. To each 0.9 ml of supernatant fluid, 0.1 ml of 8.5% sodium chloride solution was added. This procedure was suggested to us by Dr. Helen L. Casey as a modification of the W.H.O. recommendation for the treatment of anticomplementary sera with carbon dioxide [5].

Complement was prepared from guinea pigs starved overnight, etherized and bled from the heart. The serum was separated approximately 4 hours later and was stored at −70 °C.

Immune guinea pig serum, and antigens prepared from Marburg virus-infected and uninfected guinea pig livers [6] treated with 1 : 3000 beta-propiolactone were obtained through the courtesy of Dr. R. Kissling, United States National Communicable Disease Center, Atlanta, Georgia. These are referred to as the CDC serum and antigens.

Our own positive and negative antigens were prepared from vervet monkey spleen or liver, and are referred to as the PRF antigens. The majority of sera were tested with infected liver of the monkey No. D 788 described by us in another paper [4]. Tissues were crushed with carborundum, and a 20% suspension made in veronal buffer solution pH 7.4. To remove nonspecific reactive substances [7], one part of fluorocarbon (trichloro-trifluoroethane; Arcton 113, I.C.I.) was added to two parts of liver or spleen suspension; the mixture was homogenized at 4000 r.p.m. for 5 minutes and then centrifuged at 2000 r.p.m. for 10 minutes in an anglehead centrifuge. The aqueous phase was stored at −70 °C for use as antigen. Beta-propiolactone inactivation was not done, and the liver antigen from monkey D 788 had an infectious titre of 10^3 $TCID_{50}$ per ml after fluorocarbon treatment.

Complement fixation tests were carried out by the microtiter technique of Sever [8] with 2 complete haemolytic units of complement, as used by Kalter [2, 9]. Full controls were set up with each test, including back-titrations of complement in the presence of each antigen. Every serum was tested with the CDC and PRF positive and negative antigens in parallel.

Table 1 gives the approximate relation between % haemolysis and numerical values as used by KALTER [9] and followed by us.

Table 1. *Approximate Relation between % Hemolysis and Numerical Values.* (KALTER 1963)

% Hemolysis	0	30	50	75	90	100
Numerical Values		4	3	2	1	±	—

Results

1. Complement Fixation Tests

In checkerboard titrations the CDC immune serum had a titre of 1 : 128 against a 1 : 16 dilution of CDC positive antigen, which was in agreement with Kissling's own results [6]. This serum had a titre of 1 : 40 against a 1 : 64 dilution of PRF positive antigen.

The PRF immune serum showed a titre of 1 : 20 against a 1 : 32 dilution of PRF positive antigen; and had a titre of 1 : 40 against a 1 : 64 dilution of CDC positive antigen. Working titres of 1 : 16 for CDC positive antigen, and of 1 : 8 for PRF positive antigen, were used; except in the test of Marburg human sera, where a 1 : 8 dilution of both antigens was employed.

In order to establish whether we had a true Marburg virus antigen, convalescent sera from 4 Marburg patients, obtained through the courtesy of Dr. LEHMANN-GRUBE, and hyperimmune sera from two Porton guinea pigs, kindly sent to us by Dr. GORDON SMITH, were tested. The results are given in Tables 2 and 3, in which it can be seen that with each serum the titre is higher with the PRF antigen than with the CDC antigen.

Table 2. *Complement Fixation Titres of Marburg Human Convalescent Sera, Determined by* 4+ *Endpoint*

Antigens	Patients			
diluted 1 : 8	M. K.	B. O.	A. F.	H. K.
CDC +	1 : 10	1 : 10	<1 : 10*	<1 : 10***
CDC −	0	0	0	0
PRF +	1 : 20	1 : 20	<1 : 10**	1 : 40
PRF −	0	0	0	0

* 1+ at 1 : 10.
** 2+ at 1 : 10.
*** 3+ at 1 : 10.

Table 3. *Complement Fixation Titres of Porton Hyperimmune Guinea Pig Sera, Determined by 4+ Endpoint*

Antigens	Serum Dilutions	
diluted 1 : 8	G-Pig 1	G-Pig 2
CDC +	1 : 40	1 : 20
CDC −	0	0
PRF +	1 : 80	1 : 40
PRF −	0	0

For brevity in recording further results we adopted the scheme of readings shown in Table 4.

Table 4. *Recording of Results Adopted in Present Study*

Complement Fixation Reading	4+ & 3+	2+ & 1+	± & 0
Recorded Result	+	±	0

Since so few anticomplementary sera were encountered, these were not further tested; and only the results for non-anticomplementary sera are shown in the subsequent tables. Table 5 gives the findings when 268 baboon sera were tested with the 4 antigens in parallel.

Table 5. *Baboon Sera Tested by Complement Fixation*

	CDC +	CDC −	PRF +	PRF −
+	133	22	0	4
±	77	88	26	26
0	29	129	213	209

Total 268   Anticomplementary 29

No baboon sera reacted strongly to the PRF positive antigen, in contrast to the 133 strong reactors to the CDC positive antigen. Similarly, considerably more sera showed reactions to the CDC negative antigen than to the PRF negative antigen.

Table 6 presents the results for vervet monkey sera. Again there were approximately 3 times as many reactors to the CDC negative antigen as to the PRF

Table 6. *Vervet Monkey Sera Tested by Complement Fixation*

	CDC +	CDC −	PRF +	PRF −
+	159	44	7	5
±	31	63	27	29
0	7	90	163	163

Total 211   Anticomplementary 14

negative antigen; and while 159 gave strong reactions with the CDC positive antigen, only 7 reacted strongly with the PRF positive antigen.

Table 7. *Reactions (1+ or more) to Other Antigens, by CDC positive Monkeys*

CDC +	CDC −	PRF +	PRF −
159	102	33	33

Table 7 further analyses the cross reactions of the 159 CDC positive monkeys with the other antigens used. Similarly, the cross reactions of the 7 PRF positive monkeys are analysed in Table 8.

Table 8. *Reactions to Other Antigens by PRF Positive Monkeys*

Antigens	PRF +	PRF −	CDC +	CDC −
Monkey				
A	3+	3+	4+	4+
B	3+	2+	4+	3+
C	3+	1+	4+	2+
D	3+	3+	4+	4+
E	3+	±	4+	3+
F	3+	1+	4+	3+
G	3+	1+	4+	3+

The extensive cross reactions with negative liver antigens shown by these 7 PRF positive monkeys throw doubt on the validity of their positive antigen reaction; and the question of why so many animals reacted to uninfected liver antigens becomes prominent.

The sera of 8 Africans handling baboons or monkeys, and of 9 Africans employed elsewhere as cleaners in our laboratories, were then tested. Two of the animal handlers reacted strongly to the CDC positive antigen, and 2 showed trace reactions. None of the other human sera reacted with any antigen used.

The positive and negative CDC and PRF antigens were tested against the sera of rabbits immunized with Leptospira icterohaemorrhagiae, L. canicola and L. pomona. The PRF antigens showed no reactions. The CDC positive antigen gave a 2+ reaction with the L. pomona serum, and 1+ reactions with the L. canicola and L. icterohaemorrhagiae sera. The CDC negative antigen gave 3+ reactions with the L. canicola and L. icterohaemorrhagiae sera, and a 2+ reaction with the L. pomona serum.

At the suggestion of Dr. J. H. S. Gear, 14 baboon and 6 monkeys sera were tested at the South African Institute for Medical Research, for relapsing fever (S. duttoni) and syphilis antibodies. None of the baboon sera gave a positive reaction with the Kolmer complement fixation test for syphilis; but 5 monkey sera were positive, with readings of 10—40 Kolmer units. In the relapsing fever complement fixation test, 2 baboon and 4 monkey sera were positive at a dilution of 1 : 5, two of these monkey sera being positive up to dilutions of 1 : 40 and 1 : 80 respectively. Four of the 5 Kolmer positive monkeys were also positive for relapsing fever, and these included the 2 monkeys showing high titres in the latter test, suggesting a possible relationship. There was no correlation to be seen between the reactions of the 20 sera to the Marburg virus antigens and to the two antigens for relapsing fever and syphilis.

No further attempts were made to elucidate the discrepancy between the reactions of simians to the CDC and PRF antigens.

2. Virus Isolation

Since 1954 more than 3000 vervet monkeys have been used for the preparation of monolayer kidney tissue cultures in our unit, and virtually all of the tissue batches have been checked with stained coverslips. Marburg virus has not been detected in them, nor has it been encountered in previous deliberate attempts to recover viruses from South African monkeys [10] and baboons by us. In the present study, throat swabs from 84 vervet monkeys and 96 baboons were screened in tissue culture for the Marburg virus, and 235 baboon sera were similarly tested. All of these proved negative for Marburg virus, although a number of other viruses were recovered from the throat swabs.

Discussion

The report of Kalter and his colleagues [2] that a high proportion of simians gave strong complement fixing reactions to CDC positive guinea pig antigen was confirmed by us. We were, however, unable to detect convincing evidence of specific Marburg virus antibody in uninoculated baboons or monkeys when monkey tissue antigens prepared by us were used in parallel with the CDC antigens in the complement fixation test.

The occurrence of strong reactions to both guinea pig and monkey uninfected liver antigens suggests the possibility of an autoimmune state resulting from previous liver damage. The hypothesis for such a condition was first proposed by GEAR [11, 12], and has been more recently discussed by DONIACH and WALKER [13].

We did not exhaustively pursue the question of whether a second and stronger non-Marburg-virus antigen accounting for the high number of simian reactors was present in the CDC antigen. The fact that 2 African animal handlers showed strong reactions to the CDC positive antigen only, suggests that the antigen derives from a micro-organism transmissible to man.

Although the present investigation has not elicited convincing evidence of Marburg virus activity in South Africa, nevertheless the increasing use of baboons and monkeys, particularly for transplantation studies, demands that vigilance for pathogenic viruses in these animals should be intensified. Adequate facilities for the screening of primates for such viruses do not exist in most of the research centres using primates in South Africa, and their provision should be regarded as a matter of urgent importance.

Summary

Complement fixation tests performed by the microtiter technique confirmed the observation of KALTER and his colleagues that a high proportion of simian sera reacted strongly to beta-propiolactone inactivated Marburg virus guinea pig antigen prepared by the United States National Communicable Disease Center. However, when infectious antigen prepared by us from monkey liver was tested in parallel with the CDC antigen, of the 292 baboon and vervet monkey sera showing strong reactions to the CDC positive antigen, only 7 gave positive reactions to our antigen. The specificity of the reaction shown by these 7 sera is questionable in view of their cross reactions with the negative liver antigens used in the tests. Of the 292 sera reacting strongly to the CDC positive antigen, 22% also reacted strongly to the CDC negative liver antigen.

In our laboratory the use of kidney tissue cultures prepared from more than 3000 vervet monkeys since 1954, and deliberate experiments to recover viruses from baboons and vervet monkeys over a number of years, have failed to reveal Marburg virus as a natural infection. This evidence, taken in conjunction with the serological findings, indicates that Marburg virus is either not present or not very active in the baboons and vervet monkeys of South Africa. Nevertheless, in view of the increasing use of primates for research, particularly in transplantation studies, the provision of adequate facilities for the detection of pathogenic viruses in primates used in research centres is a matter of urgent importance.

Acknowledgements

We are particularly endebted to Professor R. SIEGERT and Dr. F. LEHMANN-GRUBE for the courtesy and help they have unfailingly extended to us.

We wish to thank Dr. J. H. S. GEAR, Director of the Poliomyelitis Research Foundation, for permission to conduct these studies; Dr. ROBERT KISSLING and Dr. HELEN CASEY, of the United States National Communicable Disease Center, for the generous provision of antigens and antiserum, and for advice on complement fixation procedures; Mrs. GLORIA KALTER for her guidance in our initial use of the microtiter technique; Professor H. D. BREDE and the staff of the Department of Medical Microbiology, Stellenbosch University, for the collection of baboon sera from the Cape Province; Dr. I. Purchase, of the South African National Nutrition Institute, for baboon sera from the Eastern Transvaal; and Miss J. Harding and Mr. M. Ulrich, of the South African Institute for Medical Research, for preparing the map.

References

1. Simpson, D. I. H., Bowen, E. T. W., Bright, W. F.: Lab. Anim., **2,** 75 (1968).
2. Kalter, S. S., Ratner, J. J., Heberling, R. L.: Proc. Soc. Exper. Biol. and Med., **130,** 10 (1969).
3. Dandelot, P.: Smithsonian Institution Preliminary Identification Manual for African Mammals (privately circulated) (1968).
4. Malherbe, H., Strickland-Cholmley, M.: To be published.
5. W. H. O. Expert Committee on Respiratory Virus Diseases: Technical Report Series No. 170, 40 (1959).
6. Kissling, R. E., Robinson, R. Q., Murphy, F. A., Whitfield, S. G.: Science, **160,** 888 (1968).
7. Halonen, P., Huebner, R. J., Turner, H. C.: Proc. Soc. Exper. Biol. and Med., **97,** 530 (1958).
8. Sever, J. L.: J. Immunol., **88,** 320 (1962).
9. Kalter, S. S.: Procedures for Routine Laboratory Diagnosis of Virus and Rickettsial Diseases. Burgess Pub. Co., Minneapolis: p. 39 (1963).
10. Malherbe, H., Harwin, R., Ulrich, M.: S. Afr. Med. J., **37,** 407 (1963).
11. Gear, J.: Trans. Roy. Soc. Trop. Med. and Hyg., **39,** 301 (1946).
12. Gear, J.: Acta Med. Scand., **152,** Supplement 306, p. 39 (1955).
13. Doniach, D., Walker, J. G.: Lancet **1969 1** (7599) 813.

Epizootic, Clinical, and Pathological Aspects
of Simian Hemorrhagic Fever

A. Shelokov, N.M. Tauraso, A. M. Allen, and
C. D. España

Another important recently recognized epizootic disease of monkeys is simian hemorrhagic fever. Two devastating outbreaks occurred in 1964—one in Sukhumi, USSR and the other in Bethesda, USA. Since then the disease was again recognized in California in 1967 and in Sussex, England in 1968. The clinical and pathological pictures in all outbreaks have been similar—generalized vasculitis with hematopoietic and capillary damage, resulting in a variable and diffuse localization of lesions. In all cases we were able to confirm the etiologic identity of the causal agent.

The Bethesda/64 Epizootic

Our own involvement in simian hemorrhagic fever (SHF) began in November 1964 when an explosive epizootic of febrile hemorrhagic disease occurred among monkeys held at the National Institutes of Health simian colony [1]. The outbreak occurred initially among a shipment of about 100 young rhesus monkeys which were received two weeks earlier from a new supplier in India. Besides rhesus *(M. mulatta)*, two other *Macaca* species were affected, cynomolgus *(M. irus)* and stumptailed macaque *(M. speciosa)*. Of about 1,050 animals kept in the colony at the time, 223 became ill and died of the disease during a 2 month period from November 1964 until January 1965.

The clinical characteristics of the disease at first seemed to be not unlike the illness frequently seen in the newly arrived, not yet conditioned monkeys. As a result, the earliest cases were not recognized. As more and more animals became ill, Dr. Amos Palmer, National Institutes of Health, recognized a clinical syndrome characterized by fever, anorexia, depression, tremors, facial edema and cyanosis, vomiting, melena or blood-covered hard stools, skin petechiae, retrobulbar hemorrhage, and epistaxis. The animals rapidly became dehydrated and very weak.

Typically, the animal's condition would suddenly change from apparent good health and progress rapidly to grave illness within 48 to 72 hours. Hemorrhagic lesions were not usually seen until late in the course of the disease. Skin petechiae appeared any time after 72 hours, and usually were seen only where trauma had occurred during handling. The animals survived for about ten to 15 days after onset.

Retrospective study indicated that the first death due to this disease must have occurred on November 2, 1964 in a room containing the first shipment from the new supplier. Within 2 weeks the disease had become dispersed throughout 16 of 18 rooms housing the colony and affected animals at all stages of quarantine.

Animals that were frequently handled appeared to suffer highest mortality. With the exception of only a single animal, there appeared to be no survivors among monkeys that became clinically ill.

The pathological studies have been reported [2]. Invariably we found multiple hemorrhages throughout the body, including skin, internal organs, GI tract, cranial cavity and, rather characteristically, retrobulbar space. A very striking common lesion was in the duodenum at the pylorus.

The internal organs and bone marrow were markedly pale. Characteristically the spleen was large and taut. There were perifollicular hemorrhages and peri-follicular necrosis, extreme engorgement of splenic cords with plasma, which was sometimes coagulated and replaced the blood cells usually found there. There was also follicular necrosis of the lymphatic tissues throughout the body.

In contrast with some of the human variants of epidemic hemorrhagic fever, the kidney lesions appeared non-specific. There was some smudging of the base-ment membrane, probably due to excessive protein filtration. Proximal and distal convoluted tubules usually showed cloudy swelling and fat vacuolation.

In general, the microscopic pathology was characterized by evidence of capil-lary venostasis, such as fibrin thrombi in many other organs besides the kidney, and capillary fragility, such as petechiae or gross hemorrhages into the sites of trauma.

Initial efforts to isolate the causal agent in bacteriological media, small laboratory animals and tissue cultures were unsuccessful. A hundred newly-imported rhesus monkeys were housed separately in two equal groups. The control group was held under strict isolation. Three monkeys with recognizable clinical disease were introduced as cage mates to each of 3 of the fresh monkeys in the other group. These animals were not handled in any way to cut down on the possibility of traumatic predisposition. Cage mates of the 3 affected animals and 2 animals in the adjacent cages became ill and died with the typical clinical syn-drome. However, there was not a general outbreak of the disease within the room indicating that the infection was not highly contagious. The control animals in the other isolation room remained well [1].

Next we succeeded in transmitting the disease by intramuscular inoculation of blood, serum, and suspensions of spleen and kidney from necropsied sick ani-mals. In still another set of experiments we attempted to determine the titer of the infectious agent in a specimen of whole blood which we knew was highly infectious for other animals: the titer was high, in the range of 10^{-7} to 10^{-9} per 0.5 ml inoculum.

After many unsuccessful attempts at virus isolation in a variety of animals and cell cultures, we were able to isolate a virus from 8 of the 11 serum samples. The details of virus isolation and characterization have been recorded [3].

The Sukhumi/64 Epizootic

We were not aware (until mid-1965) that a similar outbreak had occurred in August 1964 at the Soviet primate nursery in Sukhumi on the Black Sea [4, 5]. As in our case, it began two weeks after a fresh shipment of rhesus monkeys arrived from the new Indian supply house. Over a period of the next 2 weeks 28 monkeys of four *Macaca* species in different areas of the nursery became ill and died from a febrile disease. A new wave occurred in October with 34 more deaths bringing the total of cases to 62—all of them fatal.

The clinical and pathological features were essentially the same as in the Bethesda outbreak. However, the Soviet investigators at Sukhumi, predominantly neurologists and neurophysiologists, were particularly impressed with the lethargy, tremors, occasional abnormalities of the cerebrospinal fluid, and other neurologic manifestations. Suspecting encephalitis they reproduced the disease in fresh monkeys by inoculating them *intracerebrally* with *brain* emulsions, which may have further accentuated the neurologic abnormalities. We believe that these manifestations (which we had also observed in our monkeys) are a result of generalized vasculitis with variable central nervous system involvement.

The design of the Soviet studies on transmission and spread of the infection and attempted virus isolation paralleled ours with similar results. However, they were not able to isolate the virus, probably because at the time they did not have the MA-104 cell line or its equivalent on hand. Using these cell cultures we isolated the virus, Sukhumi/64, from their materials and demonstrated its identity with the Bethesda/64 virus.

Analysis of the spread of infection in the two outbreaks and the experimental studies in both countries suggested that the disease was not highly contagious. Relatively prolonged intimate contact seemed to be needed for spread from monkey-to-monkey; it was assumed that the spread was by way of infective urine. The likeliest cause of both outbreaks was carriage of the virus from room-to-room, cage-to-cage and animal-to-animal by attending personnel and research investigators, and particularly the use of inadequately sterilized syringes and needles.

It was our impression that the *Macaca* monkeys only recently encountered SHF virus in nature and that most likely some other primate or non-primate animal species was the true source of the virus. The basis for this impression was the epizootiologic evidence that all[1] monkeys with clinically recognizable SHF died and that no subclinical cases have yet occurred; besides, a survey of monkey sera indicated that monkeys lack antibodies to SHF virus. We have already noted that the first two epizootics occurred within three months of each other and could be linked to the same Indian supplier. All of these considerations led us to believe that these two epizootics were probably freak occurrences and that other SHF outbreaks were not likely, at least in the near future.

The Davis/67 Epizootic

It was surprising, therefore, to learn that another outbreak had occurred at the National Center for Primate Biology at Davis, California in October 1967. At this time, large numbers of *M. mulatta* assigned to the tuberculosis, malaria and breeding programs began to die of a disease with the clinical signs already familiar to us: anorexia, adipsia, subcutaneous bleeding at the sites of drug injections or injury, and bleeding from the nares, oral cavity, and rectum. The animals became

---

[1] The authors know of only one monkey which had clinically evident SHF and survived [1].

feverish, lethargic, and despondent; in the terminal stages occasional tremors were observed and finally coma and death ensued. The disease ran a short course with death invariably occurring within 4 to 7 days from onset of clinical signs.

At autopsy the most frequent findings consisted of hemorrhages into muscles at sites of drug injections, massive retroperitoneal blood clots, and congestion of vessels in the omentum. The spleen was usually enlarged and congested, the proximal portion of the duodenum was hemorrhagic, and the lungs were congested in some cases. There was usually loss of tone of the cardiac musculature.

The disease attacked both *M. mulatta* and *M. cynomolgus* and could be reproduced in these monkeys by inoculation of infectious serum or organ suspensions. Taking into consideration the incubation period and severity of the disease induced in *M. mulatta*, the infectious agent obtained from *M. cynomolgus* appeared to be more virulent than the one from *M. mulatta*. Whether this apparent difference was due to an increase in virulence after passage of the same agent through a different host or due to a difference in the virus *per se* is not known.

At our Bethesda laboratory we were able to reproduce the Davis/67 disease in *M. mulatta*. The clinical aspects and the pathologic features of the disease were indistinguishable from those previously noted for the Bethesda/64 [1, 2] and Sukhumi/64 [5, 6] diseases.

The Sussex/68 Epizootic

The most recent epizootics of SHF have apparently been occurring in 3 successive years in a *M. cynomolgus* colony in Sussex, England. The epizootiologic aspects of these outbreaks are currently being investigated and will be the subject of a subsequent report.

At our Bethesda laboratory we were able to reproduce the Sussex/68 disease in *M. mulatta*. The clinical and pathologic features of the disease are indistinguishable from the previously observed SHF outbreaks.

Our current knowledge of the 4 recognized epizootics of SHF is summarized in Table 1.

Table 1. *Summary of Recognized Epizootics of Simian Hemorrhagic Fever*

Date of onset	Place	Duration	No. of Primates in colony	with disease	Percent mortality*	Reference
August, 1964	Sukhumi, USSR	2 months	?	62	?	4, 5
November, 1964	Bethesda, Md. USA	3 months	1,050	223	11	1, 2, 3
October, 1967	Davis, Calif. USA	months	2,600	520	20	none
1968	Sussex, UK**	?	?	?	?	none

* Expressed as percentage of the total number of primates theoretically at risk, regardless of species susceptibility.

** A laboratory-confirmed epizootic still under investigation.

Comment

A new viral hemorrhagic disease of *Macaca* monkeys has been described. Unlike the Marburg virus disease, it is fatal for monkeys but apparently does not affect man. When first seen SHF was assumed to be of chance occurrence, but other epizootics have since been recognized. This should allert us to the possibility that subsequent outbreaks of infectious diseases of primates, especially those that are hazardous to man, may occur.

References

1. PALMER, A. E., ALLEN, A. M., TAURASO, N. M., SHELOKOV, A.: Simian Hemorrhagic Fever. I. Clinical and Epizootiologic Aspects of an Outbreak Among Quarantined Monkeys. Am. J. Trop. Med. & Hyg. **17,** 404—412 (1968).
2. ALLEN, A. M., PALMER, A. E., TAURASO, N. M., SHELOKOV, A.: Simian Hemorrhagic Fever. II. Studies in Pathology. Am. J. Trop. Med. & Hyg. **17,** 413—421 (1968).
3. TAURASO, N. M., SHELOKOV, A., PALMER, A. E., ALLEN, A. M.: Simian Hemorrhagic Fever. III. Isolation and Characterization of a Viral Agent. Am. J. Trop. Med. & Hyg. **17,** 422—431 (1968).
4. SHEVTSOVA, Z. V.: Studies on the Etiology of Simian Hemorrhagic Fever in Monkeys. Vop. Virus. **12,** 47—51 (1967).
5. LAPIN, B. A., PEKERMAN, S. M., YAKOVLEVA, L. A., DZHIKIDZE, E. K., SHEVTSOVA, Z. V., KUKSOVA, M. I., DANKO, L. V., KRILOVA, R. I., AKBROIT, E. YA., AGRABA, V. Z.: A Hemorrhagic Fever of Monkeys. Vop. Virus **12,** 168—173 (1967).
6. TAURASO, N. M., SHELOKOV, A., ALLEN, A. M., PALMER, A. E., AULISIO, C. G.: Epizootic of Simian Hemorrhagic Fever. Nature **218,** 876—877 (1968).

Simian Hemorrhagic Fever Virus

N. M. Tauraso, C. G. Aulisio, C. D. España, O. L. Wood,
and H. Liebhaber

With 3 Figures

Summary

Previous studies on the virology of the Sukhumi/64 and Bethesda/64 epizootics of simian hemorrhagic fever (SHF) are reviewed. Virologic investigation of 2 more recent epizootics (i.e. Davis/67 and Sussex/68 outbreaks) are described. Serologic studies revealed that all 4 known epizootics were caused by the same virus. Additional biophysical studies demonstrated that the buoyant density of SHF virus in $CsCl_2$ and sucrose D_2O gradients was 1.22 and 1.19, respectively. Electron microscopy of virus-infected cells revealed that unique lamellar structures precede the development of whole virions whose size corresponded to results previously obtained from Millipore filtration studies.

Introduction

In the accompanying paper was described the epizootiology, clinical features and pathology of 4 known simian hemorrhagic fever (SHF) outbreaks, namely the Sukhumi/64, Bethesda/64, Davis/67 and Sussex/68 epizootics. This report summarizes what is known of the virology of these outbreaks and describes the results of additional studies on the characterization of SHF virus.

Although SHF could easily be transmitted to monkeys of the *Macaca* species by inoculation of infectious serum or suspensions of organs obtained from sick monkeys, extreme difficulty was encountered when attempts were made to isolate the causative agent in common laboratory animals or cell cultures [1, 2, 3]. The first isolation was reported in 1968 by Tauraso et al. [2] who cultivated the Bethesda/64 agent in cell cultures. These workers isolated eight virus strains which produced similar cytopathic effect (CPE) in the MA-104 embryonic rhesus monkey kidney cell line. The three strains tested by complement-fixation (CF) and fluorescent-antibody (FAB) tests were indistinguishable. The virus causing SHF was shown to be: RNA type, less than 50 nanometers (nm) in size, chloroform-sensitive, pH 3.0—labile, and relatively heat-stable. Divalent cations, however, enhanced inactivation at 50 °C. The cell culture propagated virus produced the typical hemorrhagic disease when inoculated into rhesus monkeys *(M. mulatta)* and monkeys dying late in the course of the disease had serum antibody to the virus as determined by the CF and FAB tests.

Subsequently these same workers described the isolation in cell culture of the virus causing the Sukhumi/64 epizootic [4]. By CF and FAB tests the Sukhumi/64 virus was indistinguishable from the Bethesda/64 strains.

Materials and Methods

The materials and methods employed in most of these studies have been described in 4 earlier reports [2, 4, 5, 6].

Density gradient centrifugation. Virus-infected cells were harvested when cytopathic effect involved $75-100\%$ of the cell sheet. After sonication in a Raytheon magnetostrictive sonic oscillator at 9 kc for 3 minutes the harvests were centrifuged in an SW-25 head in the Spinco model L or L2 centrifuge at 25,000 r.p.m for 2 hours. The pellet was resuspended in a volume of tris-buffered-saline, pH 7.2 equal to 1/100th the original volume. Five to 10 ml of this 100 X concentrate was filtered through 0.2 gm celite and 1 ml of the filtrate was layered on top of a preformed $CsCl_2$ or sucrose D_2O gradient. Subsequently, centrifugation was performed using an SW-39 head of the Spinco model L centrifuge at 35,000 r.p.m. for 2 hours. Ten to 12 fractions were collected, assayed for infectivity and examined for presence of virions by electron microscopy.

Results

Virology of the Davis/67 Epizootic

At our Davis laboratory numerous unsuccessful attempts were made to isolate an etiologic agent using standard bacteriological and virological procedures, including inoculation of mice (suckling and adult), hamsters, guinea pigs, chicken embryos, rabbits, and a variety of cell cultures. The disease could be transmitted to *M. mulatta* by inoculation of blood from infected animals. Both plasma and lysed cells were centrifuged at 10,000 r.p.m. for 1 hour and filtered through Millipore filters with pore sizes of 0.45 and 0.22 microns in diameter. Transmission studies proved that the disease was due to a filterable agent present in the blood and highly pathogenic to *M. mulatta*. All animals inoculated with blood or its fractions developed a fatal hemorrhagic disease which was indistinguishable from the natural infection. The incubation period, as measured by change seen in stained thin blood smears (acid pH, stippling, intracytoplasmic granules in mononuclear cells) ranged from 3 to 6 days after inoculation and the time of death varied from 9 to 19 days, with most deaths occurring between 9 to 13 days. In addition, fecal specimens collected from sick animals and processed for inoculation, including filtration through 0.45 micron Millipore filter, were given by stomach tube to six monkeys. Two monkeys developed the disease and died 10 days post-inoculation. From these animals, samples of plasma, spinal fluid, urine and ileal contents were processed for inoculation into another four *M. mulatta*. Only the plasma and spinal fluid were infectious.

At our Bethesda laboratory we were able to reproduce the Davis/67 disease in *M. mulatta* by inoculation of infectious serum obtained from sick rhesus and cynomolgus *(M. irus)* monkeys involved in the Davis/67 outbreak. The clinical and pathologic features of the disease were indistinguishable from what was observed for the Sukhumi/64 and Bethesda/64 diseases. Although there was no difficulty in passaging the Davis/67 agent from monkey to monkey by inoculation of infectious serum, repeated attempts to isolate the virus in cell cultures have resulted in failure. However, monkeys dying late in the course of the disease were found to contain antibody to the Sukhumi/64 and Bethesda/64 virus strains.

Virology of the Sussex/68 Epizootic

Again at our Bethesda laboratory we have been able to reproduce the Sussex/68 disease in *M. mulatta* by inoculation of suspensions of organs from sick monkeys. The clinical and pathologic features of this disease were indistinguishable from that observed in the 3 previous outbreaks. The agent was easily transmissible from monkey to monkey by inoculation of infectious serum. Attempts to isolate the virus in cell cultures are currently being made. Monkeys dying late in the course of the disease were found to contain antibody to the Bethesda/64 virus.

Serologic Studies

Complement-fixation (CF) and fluorescent-antibody (FAB) tests

Attempts to identify SHF virus by serologic procedures were not successful. Table 1 lists the viral antibody reagents which were used in CF and/or FAB tests. Only one-way tests were performed between the respective antibody reagents and SHF virus 20 X concentrated cell culture antigen [2] for CF tests or SHF virus-infected coverslips for FAB tests. The only positive reactions occurred when reagents from the 4 known SHF epizootics were used. Further attempts at serologic identification of SHF virus are currently underway.

Biophysical Studies

Buoyant density of SHF virus was determined both in $CsCl_2$ and sucrose D_2O gradients [6, 8, 9]. In 4 separate experiments cell free virus sedimented to equilibrium in a $CsCl_2$ gradient at solution densities of 1.19, 1.21, 1.23 and 1.23 respectively, with an average of 1.22. In sucrose D_2O gradients infectious virus

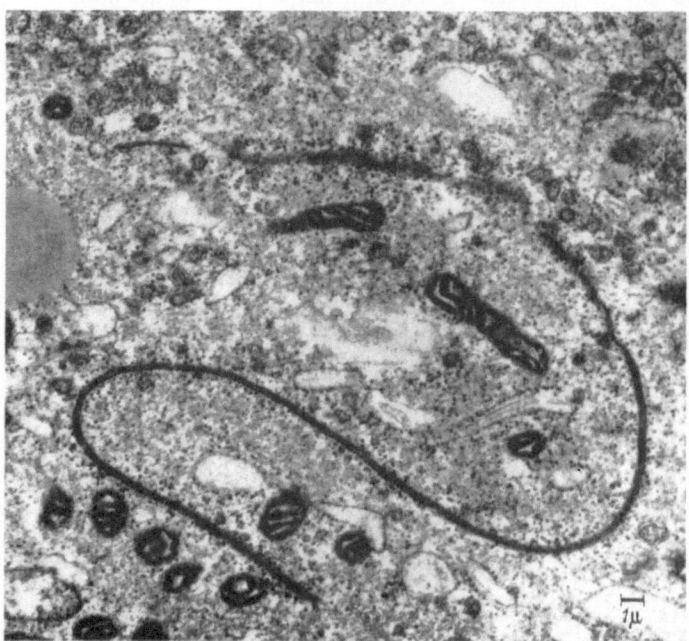

Fig. 1. MA-104 cells 48 hours after infection with simian hemorrhagic fever virus. Final mag. 49,500 ×

equilibrated at a density of 1.19 [6, 8]. The difference between the two values obtained can be expected when different substances are used to form the gradients. Both infectious virus and complete virions as seen by the electron microscope equilibrated at the densities noted above.

Electron microscopy. An electron microscopy study of SHF virus revealed some rather unique features [6]. The earliest changes in virus-infected MA-104 or BS-C-1 cells occurred as early as 24 hours after infection, at which time concentrations of electron dense material were seen in certain areas of the cell. Over the next 24 hours, what appeared to be long filamentous ribbons on cross-section developed (Fig. 1). Subsequently, double-layered forms were produced either by duplication of one or by fusion of two single-layered forms (Fig. 2). The double-

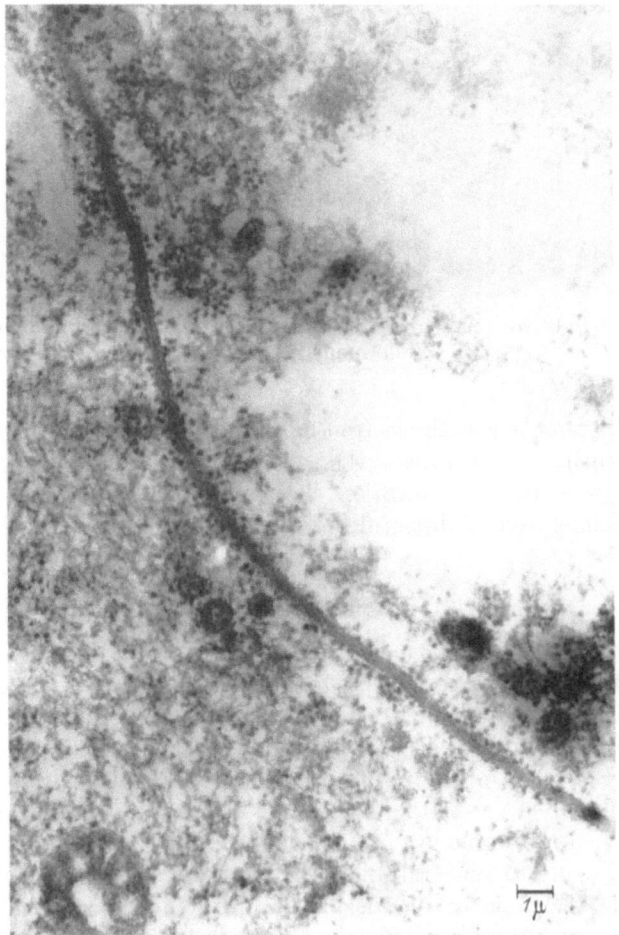

Fig. 2. MA-104 cells 72 hours after infection with simian hemorrhagic fever virus.
Final mag. 80,500 ×

layered structure was 50—60 nm in thickness and may reach a length of several microns. An examination of serial sections demonstrated that what appeared to be long filaments on cross-section were really lamellar structures [6, 8]. At approximately 72 hours after infection, the lamellar structures began to disappear and one then saw the appearance of whole virions within cytoplasmic vacuoles (Fig. 3). The final virus particle appeared to have a size of 40—45 nm and con-

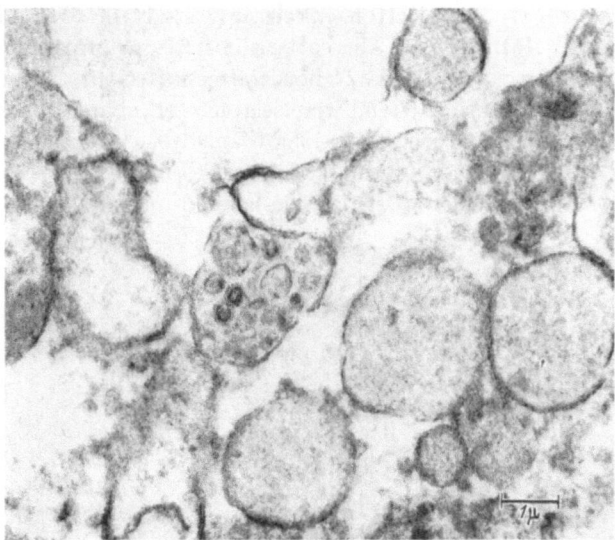

Fig. 3. MA-104 cells 96 hours after infection with simian hemorrhagic fever virus. Final mag. 123,000 ×

sisted of 22—25 nm cones with electron lucent centers. The possible relationship between the lamellar structures and the whole virions has been discussed [8]. The size of SHF virus as determined by electron microscopy agrees with results previously obtained from Millipore filtration studies [2].

Discussion

Simian hemorrhagic fever is an intriguing disease which causes a fatal disease in monkeys of the *Macaca* species [1, 3, 10]. On the basis of the epizootiologic evidence that all monkeys having clinically evident SHF die and that no sub-clinical cases are known to occur, it is felt that *Macaca* monkeys probably recently encountered SHF virus in nature and that, most likely, some other primate or non-primate animal species is the true source of the virus. Additional evidence consists of the fact that monkeys lack antibody to SHF virus. So far there have been 4 epizootics and the natural source of the virus remains unknown.

Simian hemorrhagic fever virus is also an intriguing virus. Using the present system of virus classification SHF virus might best fit into the heterogeneous arbovirus group, the biochemical and biophysical properties of which have been poorly studied. At present, there is no evidence to suggest that arthropods play a role in the transmission of SHF and, therefore, no reason to state that the virus is arthropodborne. In fact, it is known that monkeys can contract SHF from sick monkeys housed in adjacent cages in rooms known to be free of flying arthropods.

This however, does not exclude the possibility that body mites, lice or the lung mite might be vectors. Additional serology is required before excluding the possibility that SHF virus might be related to some other known virus.

Another interesting feature of SHF virus is revealed from electron microscopy of virus-infected cells. The development of lamellar structures which precede the appearance of whole virions in virus-infected cells is unique and the authors are unaware of similar structures having been implicated in the life cycle of any other virus. At this time one can only speculate on the relationship of the lamellar structures to whole virions and additional studies are needed to resolve this.

Table 1. *Results of Serologic Tests to Identify Simian Hemorrhagic Fever Virus*

Virus Group	Virus Strain	Serologic Test CF*	FAB**
ADENOVIRUS	SV-11		0
	SV-15		0
	SV-17		0
	SV-20		0
	SV-25		0
	SV-27		0
	SV-36		0
	ICH		0
ARBOVIRUS			
Group A	Bebara (MM2354)		0
	EEE (Wh. Filly)		0
	WEE (Cal 26)		0
	Una (BeAr 13136)		0
	Uruma		0
Group B	Group B, IMS		0
	Bussuquara (BeAn 4073)		0
	Dengue I (Hawaiian)		0
	II (New Guinea)		0
	III (Hammon-87)		0
	IV (H-241)		0
	V (TH-36)		0
	VI (TH Sman)		0
	Ilheus		0
	MVE		0
	Powassan (Byers)		0
	SLE		0
	WN		0
Bunyamwera	Batai (Chittoor, AMM2222)	0 (512)***	0
	Bunyamwera (Smithburn)	0 (2048)	0
	Cache Valley (Holden 6V-633)	0 (1024)	0
	Germiston	0 (1024)	0
	Guaroa (CoH 35211)	0 (1024)	0
	Ilesha (Macnam KO-2)	0 (16, 384)	0
	Kairi (TR 8900)	0 (2048)	0
	Sororoca (BeAr 32149)	0 (1024)	0
	Tensaw (A9-171b)	0 (1024)	0
	Tschalovo (P No. 5)	0	0
	Wyeomyia		0

Virus Group	Virus Strain	Serologic Test CF*	FAB**
California	CEV (BFS 283)	0 (64)	0
	Melao (TR 9375)	0 (2048)	0
	Tahyna ("92")	0 (512)	0
Miscellaneous	Anopheles A		0
	Changuinola		0
	Chagres (JW-10)		0
	EHD (SD No. 10)		0
	Icoaraci (BeAn 24262)		0
	Kemerovo (R-10)		0
	Naples SFV (prototype)		0
	Tacaribe (TR11573)		0
	Tacaribe-Junin (XJ-Parodi)	0 (2048)	0
	Tribec		0
	VSV, Ind. (BT78)	0	0
	N. J. (Lot 1522)	0	0
	Cocal (BeAr 39377)	0	0
HERPESVIRUS	H. simplex		0
	Canine herpes		0
	Simian cytomegalic		0
	B virus		0
MYXOVIRUS			
Parainfluenza	1 (Sendai)	0 (64)	0
	1	0	0
	2	0	
	3	0	
	5 (SV-5)	0	0
Miscellaneous	Avian Leukosis (RSV/RIF)		0
	(RSV-Harris)		0
	SV-41		0
	Measles	0 (32)	0
	Mumps	0 (64)	0
	Resp. Syncytial	0 (128)	0
	Rubella	0	0
	NDV		0
	NDV (Kemerovo str. 98)	0	0
	Rabies	0	0
PAPOVAVIRUS	SV-40		0
	K virus		0
PICORNAVIRUS			
Enterovirus	ECHO 1, 2, 3, 5, 6, 7, 8, 9, 11, 12 & 14		0
	SV-2		0
	SV-16		0
	SV-19		0
	Theilers GD VII		0
Poliovirus	Type 1 (Brunhilde)		0
	2 (Lansing)		0
	3 (Leon)		0
REOVIRUS	Reo 1		0
	2		0
	3		0
	SV-12		0
	SV-59		0

Table 1, cont.

Virus Group	Virus Strain	Serologic Test CF*	FAB**
Miscellaneous	Other simian viruses		
	SV-4		0
	SV-28		0
	SV-31		0
	SA-1		0
	PVM		0
	LCM		0
	Human hemorrhagic fever		0
	Crimean Type HF (Nos. 5)		0
	Korean HF (Nos. 6)		0
	Korean HF (Nos. 4)		0
	Himalaya HF (1964) (Nos. 8)		0
	Simian hemorrhagic fevers		
	Sukhumi/64	+	+
	Bethesda/64	+	+
	Davis/67		+
	Sussex/68		+
	Other hemorrhagic fevers		
	Hamster HF (Nos. 4 prs.)		0
	Human GG (Nos. 5) dil 1 : 10		0

Key to abbreviations: ICH = infectious canine hepatitis; EEE = eastern equine encephalomyelitis; WEE = western equine encephalomyelitis; MVE = Murray Valley encephalitis; WN = west Nile; CEV = California encephalitis virus; EHD = epizootic hemorrhagic disease of deer; VSV = vesicular stomatitis virus; RSV = Rous sarcoma virus; RIF = resistance-inducing factor; NDV = Newcastle disease virus; PVM = pneumonia virus of mice; LCM = lymphocytic choriomeningitis virus; HF = hemorrhagic fever; GG = gamma globulin.

  * CF = complement-fixation test.
 ** FAB = fluorescent antibody test.
*** Value within parenthesis is antibody titer to homologous virus.

References

1. SHEVTSOVA, A. V.: Studies on the etiology of hemorrhagic fever in monkeys. Vop. Virus. **12**, 47—51 (1967).
2. TAURASO, N. M., SHELOKOV, A., PALMER, A. E., ALLEN, A. M.: Simian hemorrhagic fever. III. Isolation and characterization of a viral agent. Am. J. Trop. Med. Hyg. **17**, 422—431 (1968).
3. ESPAÑA, C. D.: Personal communication.
4. TAURASO, N. M., SHELOKOV, A., ALLEN, A. M., PALMER, A. E., AULISIO, C. G.: Epizootic of simian haemorrhagic fever. Nature **218**, 876—877 (1968).
5. TAURASO, N. M., SHELOKOV, A.: Protection against Junin virus by immunization with live Tacaribe virus. Proc. Soc. Exptl. Biol. Med. **119**, 608—611 (1965).
6. WOOD, O. L., TAURASO, N. M., LIEBHABER, H.: Structure and morphogenesis of simian hemorrhagic fever virus. Bacteriol. Proceed., p. 185, 1969.
7. TAURASO, N. M., NORRIS, G. F., SORG, T. J., COOK, R. O., MYERS, M. L., TRIMMER, R.: A negative-pressure isolator for work with hazardous infectious agents in monkeys. Appl. Microbiol., **17**, 866-870 (1969)
8. WOOD, O. L., TAURASO, N. M., LIEBHABER, H.: Structure and morphogenesis of simian hemorrhagic fever virus. J. Gen. Virol., **7**, 129—136 (1970).
9. TAURASO, N. M.: Buoyant density of simian hemorrhagic fever virus in CsCl$_2$ gradient. In preparation.
10. SHELOKOV, A., TAURASO, N. M., ALLEN, A. M., ESPAÑA, C. D.: Epizootic, clinical and pathological aspects of simian hemorrhagic fever. In this book, p. 203.

Measures Taken by the Public Health Officials During the "Marburg Virus Disease"

K.-R. Nittner

In 1967 the "Marburg Virus Disease" presented the local health officials with considerable problems.

The first signs of the epidemic as recorded were:

On August 8th, 1967 an employee of the Behring Works whose job was to take care of the monkeys falls sick, suffering from fever, vomiting, and diarrhea. The illness is diagnosed as an infection of the intestines.

On August 14th four more persons are taken ill suffering from inflammation of the stomach and intestines with recurrent diarrhea. On August 15th the first patient (female) is admitted to the University Medical Clinic in Marburg in critical condition. The other four infected persons are then also hospitalized in close succession.

On August 21st the University Medical Clinic announces over the telephone that the admitted patients are suspected of having dysentery and suggests that monkeys are possibly the source of the infection. (Similar cases are reported at the Paul-Ehrlich-Institut in Frankfurt.)

On Monday, August 22nd the public health officer in Marburg begins investigating, issuing orders, etc., still under the assumption that the illness in question is dysentery.

Finally, on August 23rd suspicion is raised that the disease might be caused by a virus, since the causative organism does not respond to antibiotics.

On August 24th, 1967 the situation is as follows:

A total of 12 patients bearing symptoms of the disease have been admitted to the University Medical Clinic in Marburg. Two of the 12 have died in the meantime.

The board of health is now faced with the fact that the disease is of an infectious nature hitherto bearing the following recognized characteristics:

1. The disease is directly connected to the working with and handling of monkeys.

2. The progress of the disease appears to be malignant.

3. The causative organism as well as the manner of infection are as yet unidentified.

The health authorities must attempt to arrange for preventive measures to check the spread of the disease. These would normally include: determining the chain of infection and the source of infection (tracing back from the infected person), interrupting the chain of infection, that is, stamping out the source of infection, and adequate local investigation (including such measures as the closing of shops, offices, and schools until after the time of final disinfection). In the case of the "Marburg Virus Disease" these preventive actions were greatly impeded due to the many aspects of the disease which remained unsolved.

The public health officer was able to carry out and enforce the following necessary arrangements:

I. General hygienic action to be taken against epidemic diseases.

II. Arrangements for necessary precautions to be exercised in the Behring Works.

III. Internal measures to be enforced by the public health officials.

I. Overall Hygienic Measures

All instructions and general preventive measures were from the first carried out in collaboration with the University Medical Clinic, the Institute of Hygienics in Marburg and the Behring Works. This arrangement proved especially efficient and advantageous during the entire course of the epidemic.

The following detailed measures for hygienic epidemic prevention were put into action:

1. Every practising physician received a pamphlet familiarizing him with the initial symptoms of the disease as far as these had been determined. The pamphlet placed suspicion upon a virus infection originating in the tropics. It also pointed out that the blood of an infected person as well as the blood and body tissue of monkeys could prove highly infectious. Every physician was asked to call the public health office immediately if a patient's symptoms appeared suspicious. All doctors were urgently requested *not* to hospitalize anyone who bore the symptoms of the disease without *first* notifying the health authorities.

2. All suspicious cases were first examined by a public health commissioner in order to prevent unnecessary hospitalization. If the diagnosis was not clear enough, his medical colleagues from the University Medical Clinic were asked for aid in reaching a diagnosis. These precautions proved very successful. During the entire epidemic not a single suspicious case was hospitalized that later proved to be entirely unrelated to the disease in question. On the other hand, persons sick with the "Marburg Virus Disease" did not go undetected in their homes. Only those patients were admitted to hospital who had previously been cleared by the health office in agreement with the University Medical Clinic.

3. Since the exact manner of infection still remained unknown and, theoretically, any number of possibilities had to be taken into account, all those persons who had been in first-degree contact with the disease carriers (sometimes amounting to as many as 100 persons) were placed under observation. These people were visited and examined daily by a physician from the public health office.

4. In accordance with § 37, Section 1 and in connection with § 35, Section 1 of the Federal Law passed on July 18, 1961 regarding epidemics (BSG), all persons infected with the disease were to be submitted to hospital care. All those sick or suspected of having the disease and all persons suspected of infection were ordered to be hospitalized. The head physician was to determine the necessary period of isolation.

5. The following general restrictions were placed on those suspected of being disease carriers:

1. They were not allowed to prepare food for public consumption.
2. They were forbidden to attend school.
3. They were not allowed to donate blood.

The first two restrictions were removed as soon as the manner of infection was specifically clarified and it was certain that the disease was not communicated by means of the faeces or urine. The restriction placed upon blood donation continued.

II. Precautions Exercised in the Behring Works

(These measures were also recommended and approved by the Behring Works)

In their conference on August 18th, 1967, the managers of the various departments of the Behring Works came to the conclusion that the disease was apparently a group disease (bacterial intestinal infection). Thus on August 19th and August 22nd, a restriction was placed on all operative work dealing with monkeys. Those persons charged with the care of these animals were required to wear gloves and sterilized face masks. On August 25th all monkeys and those animals harbored in the same buildings with the monkeys were put to death. On August 26th the carcasses were burned using the proper cautionary measures. The final disinfection was carried out on August 28th.

In addition, the public health office also ordered the Behring Works to notify them immediately in case of any new outbreaks of the disease.

III. Measures Taken within the Public Health Administration

1. Under this point we particularly wish to take into consideration the daily consultations between the public health commissioners and the medical representatives from the University Medical Clinic, the Institute of Hygienics in Marburg and the Behring Works—especially at the outbreak of the epidemic. As mentioned previously, this cooperation proved extremely successful in combating the epidemic.

2. Press conferences were held after each meeting. These also proved very favorable, since they helped to prevent a general panic and aided in the instruction of the public.

3. It was also necessary to discuss plans for a quarantine station. Not only would isolation rooms be necessary, but these rooms would also have to be adequately furnished and nurses would have to be available to attend to the sick. The fact that we would have been short-handed was quite certain, since no one can be forced to attend to a patient who is under quarantine. It would have proved difficult to set up a quarantine station in other respects as well.

4. All precautionary measures and orders were constantly reviewed and checked with those being simultaneously employed in Frankfurt. Before the local health officials could undertake anything, their suggestions had to be approved by the proper district officials (mayor, territorial district advisor = "Landrat") after they had been duly notified and instructed. The approval and cooperation of the

"Regierungspräsident" (district administrative president) in Kassel as well as the Ministry of Health, Labor, and Welfare in Wiesbaden was also required.

During this time the public health officials had to deal with many other special measures in addition to the above-mentioned items. For example, many questions arose concerning the suitability of the city of Marburg as a vacation resort. East German citizens who had been travelling in Western Germany had to be issued special certificates before East German officials would allow them to return across the border. The public prosecutor's office intervened in a prosecution case against a person unknown. Questions pertaining to the disinfection of sewage also had to be examined. (The question of sewage disinfection arose from a "false alarm" sickness.) Autopsy problems had to be regulated, such as the transportation of corpses, correct burial procedures, etc. (inquest to be held by the public health commissioner, suitable disposal of all devices used for the burial, etc.). All these precautionary measures were in keeping with the regulations proposed by the "Regierungspräsident" in Kassel. Last of all there was the disagreeable matter of taking care of the burial costs.

I may note that upon inquiry the public health office in Marburg was able to inform the Pasteur Institute in Paris that the epidemic in question was not of the same nature as a plague.

To summarize, the following conclusions may be drawn from experiences gained during the "Marburg Virus Disease" incident:

1. In dealing with infectious diseases a prompt and detailed report of the disease is required in order to adopt hygienic methods of combating epidemics.

2. Quarantine stations—equipped with adequate stock and the necessary personnel—must be planned and provided for in such a manner that they can be put into service in a very short period of time. This appears practically impossible except when carried out on a district administrative level.

3. The setting up of a central virus laboratory is urgently required.

4. Federal legislative regulation is necessary in connection with the importation and the maintenance of monkeys.

In view of these requirements, these were the positive findings:

I. The prevention of large-scale epidemics can only prove successful if there is good cooperation between the Medical Clinic, the Institute of Hygiene (possibly veterinary officials) and the practicing physicians on one hand and the local health authorities on the other. Not only joint theoretical considerations and planning are necessary but also practical measures (i.e. first examination of suspected cases etc.).

II. Prompt and purposeful action is necessary. On the one hand, panic is to be avoided, on the other hand definite rules and restrictions must be established.

III. The legal provisions as expressed in the Federal legislation on epidemics are adequate if handled in a flexible manner.

In conclusion, I may say that the "Marburg Virus Disease" provided the public health officials with new insight and knowledge which can and should be utilized in similar matters concerning epidemics and general hygiene.

The Vervet Monkey Disease
Protection Against Occupational Hazards

W. ANDERS

In the question of protection against hazards arising from imported monkeys, we have to differentiate between importation and spreading of the disease. We have to check whether one or both of these events can be influenced by the action of Public Health Authorities.

There are rules distributed by our "Berufsgenossenschaften" (that means bodies specially set up by the government to cover prevention of professional accidents, including occupational diseases, and payment of benefits).

The analysis of the epidemiology of the Marburg disease leads us to some aspects important enough to call for an answer to the above mentioned question.

Following up the route of a vervet from capture in the bush to the operating theatre, we may distinguish four sections —

1. Capture and keeping before dispatch;
2. Transport;
3. Quarantine of the imported monkeys;
4. Scientific use of the animals.

Of these, only items 3 and 4 can be made subject to regulations by German authorities, whereas for items 1 and 2 there can only be recommendations. However, the recommendations for item 2 are prerequisites for granting exceptional import authorizations. Such exceptional authorizations may be granted by the government of the Federal "Land" under whose jurisdiction the airport of destination will come.

Both regulations and recommendations do not only consider the aspect of general accident prevention as mentioned above, but also provide for protection against accidents in vaccine production.

1. Importers are requested to see that the monkeys

a) are taken from a defined habitat and not from different regions of a country;

b) are in apparent good health as a population. This can be ascertained by watching for any signs of mass-dying among wildlife animals;

c) should not be provided with "Health Certificates" because these cannot be true. It would be worthwhile to think about the possibility of establishing direct surveillance in the natural habitat to be carried out by veterinarians who might be commissioned by WHO.

2. Another aspect of granting exceptional import authorizations will be that the importation be carried out by transport on a direct route without transhipment, since any contact with other animals must be avoided.

On the basis of experience, it is recommended that

a) a shipment should not exceed more than about 200 animals in number;

b) the cages must be of a size to give sufficient room to the animals yet small enough to avoid injuries to the animals by turbulences in flight;

c) the animals should not be caged more than 6 to a box in individual compartments;

d) there will be neither food nor drink for the animals to avoid soiling of the whole consignment.

Even the best recommendations, however, are not capable of relieving the animals from the "stress". Investigations of the Marburg incident revealed that meteorological conditions during the flight of the shipment of monkeys carrying the disease were such that the monkeys both suffered a stress condition and body injuries. There were considerable turbulences over southern Europe extending up to the north of France and particularly heavy ones over the Alps.

Onward transportation from the airport to the final destination is to be carried out in accordance with special requirements laid down in the import authorization.

3. Quarantine: The guidelines for the prevention of infections of man by monkeys in cases where the importation is permitted, requires at least 12 weeks of surveillance in a closed environment, covering the monkeys not only as individuals but also as a group. The import authorization will require 6 of the 12 weeks to be under official surveillance.

During the quarantine period, the health of the animals should be checked by an experienced person. Such person is not required to be a veterinary surgeon.

Monkeys from one and the same consignment must be housed together in one unit, and the period of observation must end at the same time for all of them.

At the end of the quarantine period, a specially designated officer called "Beauftragter für den Infektionsschutz" (Commissioner for the protection against infections) has to give his approval for the intended use of the animals.

Even then, the animal can be kept outside the confined unit only if special precautions are observed.

There will be no detailed requirements for the construction of a quarantine unit; this might render the importation of monkeys virtually impossible. Existing general requirements concern ventilation, cages, and general layout of the rooms. A lock with shower and facilities for a complete change of clothes are imperative.

Construction alone does not warrant enough safety. The care of animals is a highly responsible job, for which well-trained persons are needed. In Germany, recognition of this fact has led, as a first step, to the adoption of "animal technician" as a profession. Since 1967 an "animal technician" certificate (German: Tierpfleger) may be granted on the basis of an examination following 3 years of special training. This implies also the achievement of an adequate pay structure. It is to be hoped that in the future there will be a development towards an "animal technician" status comparable to that in Britain.

4. All that has been outlined for the quarantine period also applies for the time the animals are used. Contact with the animals' blood which is extremely

dangerous can be avoided only by working under strictly aseptic conditions. Such conditions must be observed by anyone who might have contact with animal material.

Implementation of the rules for protection is the task of the already mentioned "commissioner for the protection against infections" who is to be appointed by the employer, irrespective of whether the latter represents a private or national body.

Finally, there is also the problem of medical care for those who might be victims of an accident. Naturally, the patient must be isolated. Here too, a hazard will arise for the staff who have to handle the patients' blood in any way. The disease may be transmitted from man to man, particularly through contaminated needles penetrating protective gloves.

In conclusion, it can be said that observation of what has been outlined for capture and transportation is likely to reduce the risk associated with the importation of monkeys.

Strict adherence to the rules concerning quarantine and the handling of monkeys and monkey material in experimentation is capable of preventing a spread of the imported disease inside Germany. Even after quarantine, there will always remain a risk of infection associated with monkey material. There must be a complex of measures covering the whole lifetime of a monkey to diminish the risk to those handling them.

Legislative Measures Concerning the Importation of Monkeys

W. Schumacher

In view of the unexpected and rather dramatic incidents of disease in Marburg and Frankfurt, the question arose as to whether legal measures were required in order to prevent a repetition of such an occurrence. Until then no such requirement had arisen, since, in spite of a considerable import of monkeys during the two last decades, serious incidents had never been reported. On the other hand, a survey made when the laboratory staff became ill revealed that the imported monkeys were destined not only for scientific institutions and zoological gardens, but to an astonishing extent also, by way of trade to private persons. As the incidents of disease, however, were a clear warning of the dangers which may —even though perhaps only in the case of a concentration of unfortunate events— emanate from such animals—a general prohibition of the import and transit of monkeys (Simiae) and lemurs (Prosimiae) was decided on. Exempted therefrom are only animals which are used in circuses. To meet the requirements of scientific institutions, vaccine producers and others the competent authorities may, in agreement with the Federal Minister for Food, Agriculture and Forestry, grant exemptions from the import prohibition, if an importation of diseases need not be apprehended. The relevant special licences may be granted subject to conditions and impositions. Thus, control of the import and of the whereabouts of the monkeys is ensured. Special licences are, as a rule, only issued to scientifically controlled institutions, inclusive of the establishments for vaccine production, for the purposes of carrying out scientific tests, vaccine production and examination, and for producing biological material, as well as for German zoological gardens which are considered as nonprofit institutions. The destination of the animals for the mentioned institutions must already be proved by the importer when making the application.

The conditions prescribed in this connection refer to the transport of the animals, to a veterinary check at the frontier as well as to a quarantine prior to their use. I will now mention only the most essential points and wish to emphasize, above all, that the measures concerned are not rigid regulations, but recommendations which may be modified by the responsible veterinary officer in accordance with the conditions of the individual case.

The transport of the animals should be carried out only by aeroplanes in direct flight; import by sea may be allowed in individual cases, in particular if monkeys are destined for zoological gardens. Necessary transfer for onward transmission by air to another airport in the Federal Republic of Germany or in a foreign country may be permitted in a specific case, provided this is not opposed for veterinary reasons.

The transport of monkeys with other animals in one and the same freight space of an aeroplane is prohibited. Transport cages to be used repeatedly must be easy to disinfect and of a design such as not to hurt the animals (e.g. without sharp corners or edges etc.).

On their way from the country of origin to the consignee, the animals must be conveyed singly or in pairs in the transport cages. For certain species of monkeys who, due to their sensibility, would suffer in single transport (e.g. rhesus monkeys) exemption may be granted—but not for more than 10 animals in one transport cage.

As a single cage in which the animals must be transported singly or in pairs, there may be permitted also large cages, so far as these are adequately subdivided and the regulation on accommodation of up to 2 animals in one partition is observed.

The importer has to make arrangements that, on arrival of the aeroplane, a suitable transport vehicle be available, into which the transport cages must be transferred directly from the aeroplane. The conditions of the transport vehicle must be such as to prevent any leakage or dropping of animal excrements, litter and feed. In addition thereto, the freight space must be easy to disinfect. From the airport, the animals must be conveyed directly to the consignee without any further transferring or additional loading. The simultaneous conveyance of other animals and goods is prohibited.

On entry at the frontier, the animals are subjected to a veterinary check. At the border station, there must be placed before the official veterinarian, competent for the transfer station, certificates stating that the animals were examined directly prior to their transfer into the aeroplane and that their general health condition was found to be good and that they had not shown any symptoms of a communicable disease.

During the veterinary check at the frontier, the transport cages should, if possible, not be opened. Dead and diseased animals remain in the transport cages. Experience has proved that, where the transport of monkeys is concerned, deaths must frequently be reckoned with, without any communicable diseases being the cause, so far as this has hitherto come to our knowledge. If, from the number of the dead and diseased animals or from the behaviour of the animals, the presence of a communicable disease may be assumed, the consignment must be refused. In the case of its return not being possible, all monkeys must be killed and disposed of with due precautions.

The veterinarian at the frontier notifies the veterinary officer responsible for the place of destination by cable of the number of animals in the transport and of the date of the onward transmission, as well as of special details, e.g. cases of disease and death in the consignment.

The veterinary officer has then to ascertain whether the consignment corresponds to the data given in the accompanying documents. Furthermore, he examines the general condition of the animals.

At the place of destination, the monkeys must be kept under official observation during a period of 6 weeks. The official observation is deemed to be withdrawn if, at the final examination by the veterinary officer, the animals prove to be not suspected of disease. In the case of monkeys which are introduced into German zoos from other zoological gardens, where, according to evidence, they had been for at least 8 weeks, the official observation period may be dispensed with.

During the period of observation, it must be ascertained that, by proper separation, contact of the introduced animals with any other animals is impossible. In zoological gardens, separation in the normal animal houses is considered adequate.

The competent authority may permit that the official period of observation to be reduced or waived, if the type of the scientific tests or work so requires and if, having regard to the circumstances of an individual case, the competent health authority deem this to be justifiable.

With regard to the import of monkeys and lemurs in tourist traffic, no particular hazard of contamination need be feared; the monkey brought in by the passenger and owner of the animal should be kept merely in the personal care of the latter. The animal must neither be sold, nor given away as a present, and must be exported again on departure.

Any disease which an animal may develop within 3 months subsequent to its having been imported must be notified immediately to the competent veterinary officer.

Even though these regulations aim at the greatest possible extent of safety, they are, nevertheless, flexible enough to take into account the requirements of science and practice. When weighing the pros and cons of an individual case, the protection of man, in particular of the staff who have to deal with the animals, must, of course, be paramount.

W.H.O. Draft Recommendations for the Supply, Safe-Handling, and Use of Non-Human Primates for Biomedical Purposes

W. I. B. BEVERIDGE

With the purpose of improving the health status of primates available for laboratory work in the future, WHO has drafted recommendations dealing with the capture, export, import, and handling of these animals. The essentials of the document, which is only in draft form, may be summarised as follows:

1. The animals should only be caught in areas and at times of the year likely to yield healthy specimens and they should be given a permanent identification mark as soon as caught. Ordinarily, they should be quarantined under suitable circumstances for three weeks before export, but exceptions may be made when there are special requirements and the animals may then be taken straight from the trap to the airport. The intention is to avoid holding numbers of animals under conditions of poor hygiene, as often happens at present.

2. Transportation from exporting to importing country should be as rapid as possible, preferably by jet aircraft without intermediate stops. During transport each animal should be kept separate in a suitable container.

3. The importing country should quarantine the animals for 6 weeks under suitable conditions, caged singly or in pairs. Different shipments and different species should not be kept in the same room. Special mention is made of TB tests and X-rays, and certain other precautionary measures concerning communicable diseases. Personnel should wear protective clothing, be examined regularly for TB and should observe appropriate precautions.

4. After the animals have been released from quarantine, all personnel associated with either them or tissues derived from them should be warned of possible dangers and take appropriate precautions.

5. It is recommended that national veterinary authorities in the country of origin should supervise capture, quarantine, and export; also in the importing country the national veterinary authorities should supervise import and quarantine. National public health authorities should ensure that adequate precautions are taken in laboratories where primates or their tissues are used.

These draft recommendations have been widely distributed for comment. The majority of replies have been favourable and indicate a willingness to accept the recommendations without amendment. However, there are a number of replies which suggest that changes should be made in the document. Probably WHO will call in some consultants to advise on redrafting, and the revised document will be placed before a W.H.O. Scientific Group in September 1970.

Marburg Virus: Consequences for the Manufacture and Control of Virus Vaccines

O. BONIN

The consequences of the outbreak of hemorrhagic monkey fever for the manufacture and control of virus vaccines may be seen under several aspects of a more limited or a more general importance. I shall not dwell upon some transient consequences restricted to the Behringwerke and the Paul-Ehrlich-Institute where the disease occurred, although they caused considerable difficulties in the continuation and expansion of both production and control of several vaccines. Most of these difficulties, fortunately, could be solved within a fairly short priod—at least, as far as Behringwerke are concerned. I shall discuss, however, two general aspects of a more permanent importance. At first a few words must be said about the choice of cell cultures for vaccine production—although the tendency to shift from monkey kidney cultures to other cell systems did not begin with this accident. And finally, we have to deal with some safety precautions for vaccine production and with practicable methods for detecting this particular agent when present in individual vaccine lots.

Starting with the first point I should like to express clearly our opinion that the isolation of a new agent from monkey kidney cultures is not an adequate reason for banning this cell system from vaccine production—even if the agent is as dangerous as Marburg virus. Retrospectively it seems understandable that some of the early reactions to the accidents in our country definitely overshot the mark. Thus, what Dr. HILLEMAN assumed at the conference on cell cultures in virus vaccine production — that the monkey kidney cell system would be abandoned — has not been confirmed by the German experience. According to our present knowledge, the Marburg virus is not more dangerous or mysterious than for instance, B-virus is. Dr. STONES certainly was right when he stated in the same discussion in Washington that we are used to living with B-virus for years—why should we not get used to living with Marburg virus too? Since we know how to detect the agent and how to protect vaccines and personnel from it, the Marburg virus is only one item in a list of more than 50 simian viruses—although one of the more dangerous ones.

The objections against the monkey kidney cell system for vaccine production arising from the simian viruses are more of a practical nature than of a general one. Many of these viruses require special tests for their detection. The longer the list of simian viruses becomes, the more tests must be performed with each lot of monkey kidney cells used for vaccine production. And the more viruses we have to exclude, the higher becomes the percentage of cell lots which must be discarded because of contamination.

For manufacture of some newer virus vaccines, quite a few other primary tissue culture systems are in use instead of monkey kidney cells. Measles virus according to the U.S. requirements may be grown in canine kidney or chick embryo cells.   Rubella vaccines are produced in rabbit kidney, canine kidney, and duck embryo cell cultures.   The latter culture system seems especially interesting in connection with our question.   This refers to the HPV 77 strain which was

Cell cultures suitable for manufacture of some vaccines against viral diseases of man

Vaccine		Primary cell cultures prepared from						Cell cultures grown in serial passage HDCS
		monkey kidney	rabbit kidney	canine kidney	calf kidney	chick embryo	duck embryo	
Poliomye-	inact.	+						+
litis,	live	+						+
Measles,	inact.	+		+		+		+
	live			+		+		+
Adeno-	inact.	+						+
virus,	live	+						+
Rubella,	live	+	+	+			+	+
Smallpox,	inact.	+	+		+			+

attenuated by Dr. PARKMAN in monkey kidney cells.   For vaccine production it was adapted to duck embryo cells in order to eliminate the frequently contaminated monkey kidney cell system.

Difficulties arise, however, in the replacement of monkey kidney cells by other primary cell systems in manufacture of poliovirus vaccines, either inactivated or live.   These viruses multiply only in cell cultures originating from primates. Since primary human cells are very difficult to obtain in large amounts and since such cells might be even more dangerously contaminated—for instance by hepatitis viruses—the monkey kidney culture remains the only primary cell system suitable for poliovirus vaccines.

With the exception of monkey kidney cells, such vaccines can only be produced in human fibroblasts grown in serial passage which maintain their diploid chromosome pattern throughout their limited life span.   After extensive laboratory testing and clinical trials the process of licensing such vaccines for commercial manufacture is in progress now in many countries.   For Germany, a license for production already has been issued, and the requirements for control have been formulated and are under discussion at present.   Thus we hope to have vaccines available soon for which the Marburg virus is no longer a problem.

I must repeat, however, that this development has not been induced—not even strongly reinforced—by the detection of the Marburg virus.   It was started already, long before we knew about this agent.   Now the Marburg virus is only one stone in the mosaic picture we have about monkey kidney cultures—that they are very frequently contaminated by extraneous agents and therefore cause many problems in vaccine production.

Human diploid cells within a reasonable time span certainly will not replace the primary monkey kidney cell system in manufacture of all virus vaccines.   At least some inactivated vaccines will continue to be produced in monkey kidney cells since the supply with WI 38 cells—the only continuously propagated and sufficiently tested cell system—is limited and since it will take several years to accumulate equivalent experience for other diploid cell lines.

Adequate safety precautions, therefore, must be introduced into the process of manufacture of vaccines grown in monkey kidney cells and adequate tests must be performed for such vaccines in order to demonstrate that they are not contaminated by the virus of hemorrhagic monkey disease.

Such safety precautions and such additional tests should be required not only for vaccines manufactured in cells from African monkeys. They should apply equally to products grown in cells from any other monkey species. Although the disease has been imported to Europe by African monkeys, the origin of the agent remains uncertain. Studies performed by the groups of Dr. HAAS and Dr. GORDON SMITH have demonstrated that monkeys of Indian origin are equally susceptible to the agent. At least theoretically, every monkey might be a carrier of the Marburg virus and transmit it to cell cultures for vaccine production.

The most important safety precaution for manufacture of vaccines certainly is a proper quarantine for all monkeys used for preparation of cell cultures. A minimum quarantine period of 6 weeks is required by most of the national and international regulations. A code of conduct for the safe handling of nonhuman primates, elaborated recently by Dr. PERKINS and approved by an international group at a meeting in Brighton, recommends a prolonged quarantine period of at least 12 weeks—as do the regulations of German occupational insurance. If any infected monkeys are introduced into a group, they most probably will die during such quarantine. Conditioning the monkeys for such a period, however, does not completely exclude the spread of infection within the group if latent infections in animals occur or if, exceptionally, single animals become excreters after recovery.

With respect to further safety precautions during the process of manufacture, it has been claimed that a virus as big as Marburg virus should be eliminated from vaccines or tissue culture fluids by the usual filtration procedures. Filtration certainly reduces the virus content of such fluids and in parallel also the risk for the vaccines. One cannot rely on this, however, since failures may occur and since a few infective particles may enter vaccines through faulty filter pads. Virus suspensions for vaccine production, therefore, should be tested for Marburg virus before filtration. If they pass such tests, the filtration procedure is an additional safety measure to prevent the issue of infected vaccines.

For the detection of Marburg virus in such materials, two methods of about equal sensitivity are available. The agent can either be detected by inoculation into susceptible tissue cultures and subsequent immunofluorescent staining, or it can be isolated by inoculation into guinea pigs.

For the new German requirements for the oral poliovirus vaccine, we have agreed to introduce an additional guinea pig test and that for two reasons:

1. Tissue culture tests involving large sample volumes are very laborious in the stage of immunofluorescent staining. In addition they require large volumes of labelled antisera which are not easy to obtain. And finally they must be checked carefully by positive and negative controls in order to avoid misinterpretation.

2. Most of the tissue culture systems suitable for detection of the Marburg virus are also sensitive to poliovirus. For the control of such vaccines the poliovirus would have to be neutralized by high titer antisera tested for ab-

sence of antibodies against the Marburg agent. This would require continuous work with this agent in the control laboratory, which seems dangerous not only for the personnel but also with respect to cross contaminations.

The guinea pig test must be performed under adequate safety precautions. But it does not require continuous work with the active virus—either in sensitivity tests or in the evaluation of hyperimmune sera. It seems to us therefore more suitable for control purposes.

The guinea pig test, already included in many regulations for individual biological products, does not seem to be optimal for the detection of Marburg agent. Since it was designed for recognition of tuberculosis, the body temperature of the animals is recorded only between the 4th and the 6th week of the test. In the new German regulations mentioned already, we have required that the temperature is measured all over the 6 weeks observation period.

Raised temperature alone, however, is not a reliable criterion for the decision whether an animal is infected or not. We have seen quite a few guinea pigs presenting typical histopathology without any previous fever. On the other hand, sometimes a transient fever occurs without any typical pathology. For a reliable test we therefore must require that the animals are examined histopathologically or that at least one blind passage is performed.

Our new regulations for poliovirus vaccines therefore will require that, in addition to the old test, five more animals are injected intraperitoneally with the vaccine under test. These animals are controlled for body temperature for a period of two to three weeks. After this period a blind passage using blood and liver or spleen suspension is done into 5 animals again. Guinea pigs showing fever or other signs of disease are examined separately, including subinoculations of the same materials.

We feel that this test, which is very similar to that used in Canada, is a useful tool for detecting the Marburg agent if present in vaccines.

In this respect our new regulations are more strict than those in the U.S.A. where the additional guinea pig test was rescinded in June 1968. As long as we know so little about the origin of the disease, and as long as we are not absolutely sure that such events will never recur, we are of the opinion that no vaccine lot manufactured in monkey cells should be released for the market unless it is controlled in this or a similar test for the absence of this dangerous agent.